# AROUND THE WORLD

## A CELEBRATION OF CIRCUMNAVIGATION

# AROUND THE WORLD

# CONT

# ENTS

Clockwise from top:
American aviators and
a Douglas World Cruiser
in 1924; motorcyclist
Anne-France Dautheville;
a replica of Magellan's
Victoria sails up Spain's
Guadalquivir river in 2019.

# INTRODUCTION

If you thought that a journey all the way around the world would yield some incredible stories, you'd be right. When our authors were researching this book in libraries and archives, two things became apparent. One was that there had been far more extraordinary attempts – mostly successful – than we had expected. And, secondly, the journeys had been made by otherwise ordinary people. There was Jerrie Mock, an Ohio homemaker who declared: 'If I don't get out of this house, I'll go nuts'. She set off in her single-engined Cessna in 1964 and was the first woman to fly solo around the world. And there was my great-aunt Mavis, who drove a Land Rover to Australia with four other grandmothers in 1968, via the Khyber Pass, and circled back by ship. In the early 1970s, Anne-France Dautheville departed Paris on her 100cc Kawasaki, on her way to becoming the first woman to motorbike around the world. 'I am not exceptional at all', she told the New York Times. Of course, there is also a fair share of chancers and glory-hunters among the circumnavigators.

The first people that we know to have completed a circumnavigation were the Portuguese explorer Ferdinand Magellan and some of his crew (very few of the original 237 men made it back). They had ventured across the oceans 500 years ago in five ships, only one of which, the Victoria, returned to Spain in 1522. In those days the world as a whole was largely unknown, except to the inhabitants of each region. Today, we can see the planet in detail on Google Earth, yet in many ways parts of it are less accessible than they were 50 or 500 years ago. For this book, we dedicated each chapter to one of nine modes of transport that people have taken around the world – from bicycles to balloons – and focused on one particular story for each chapter. The gatefold maps for these in-depth accounts are created to reflect the world of that period.

The most direct way to circle the globe, if one was to follow the Equator, would take about 25,000 miles (40,000km). But most people choose a more meandering or a more efficient route. Some try to complete a circumnavigation as quickly as possible (there are rules and regulations to follow if you wish to claim a record). Others spend a lifetime on the road. Whatever your preference, we guarantee that by the time you finish reading these amazing tales, you'll be considering a circumnavigation of your own.

MOTORCYCLE: Ted Simon p266

PLANE: United States Army Air Service p76

BOAT: Francis Chichester p102

TRAIN: Nellie Bly p230

BALLOON: Graf Zeppelin p198

SHIP: Magellan–Elcano p12

BICYCLE: Mark Beaumont p42

CAR: Aloha Wanderwell p168

# 500 YEARS OF CIRCUMNAVIGATIONS

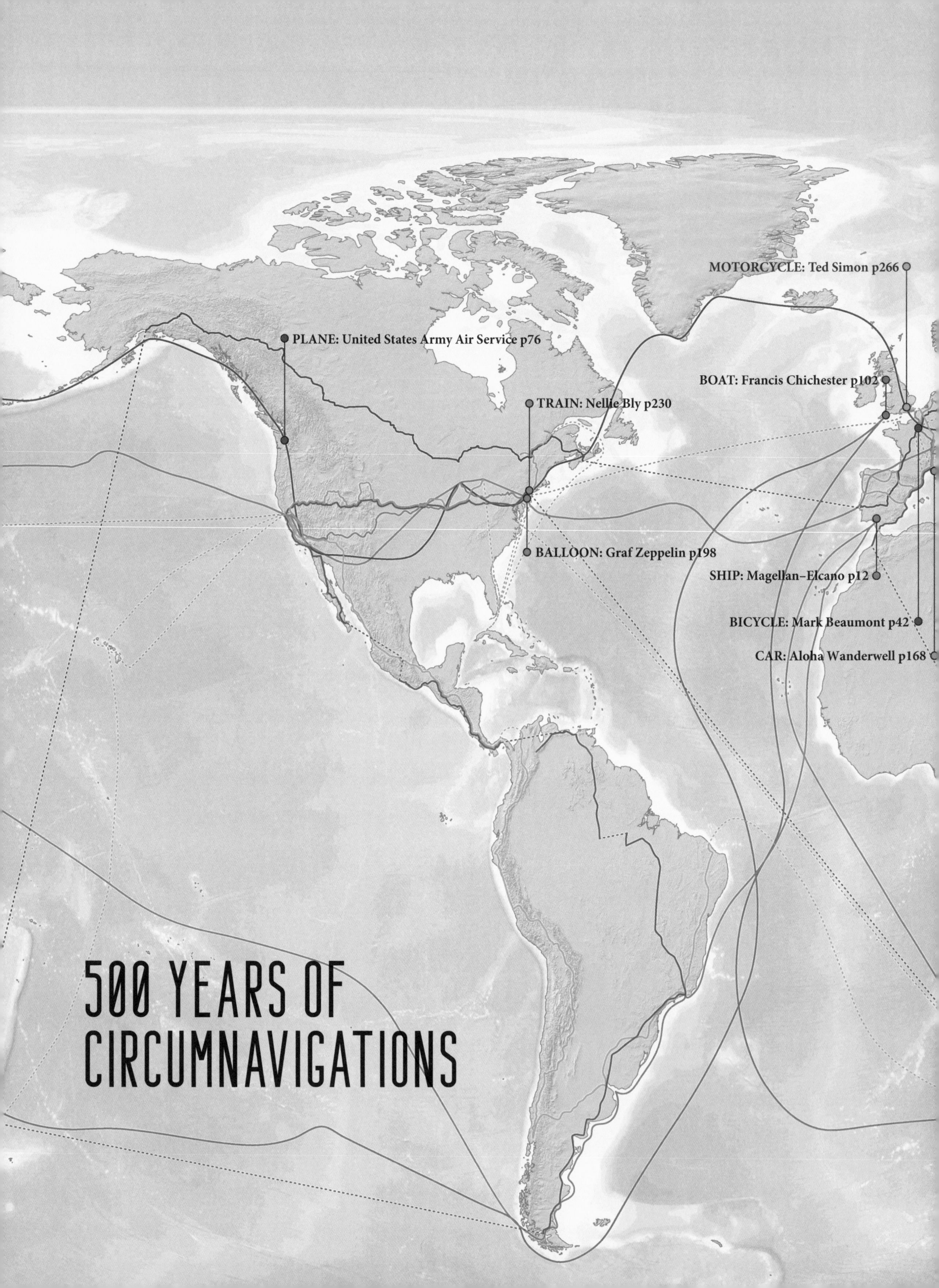

FOOT: Robert Garside p136

# SHIP
## BY

I t might seem trivial that the first circumnavigation of the world was motivated by food – and condiments at that – but bear in mind that, in the 15th century, the 'spice routes' between Europe and Asia were as much about maritime power, and in particular the ambition to establish trade monopolies, as they were about the menus of the great and the good.

Spices such as clove, nutmeg, cinnamon and mace were premium products, and the particular tropical conditions required to grow them demanded, at least for Europeans, a considerable sea voyage. The expense of importing these spices in large quantities made them a highly tradeable commodity; finding the quickest route to the Spice Islands, that small band northeast of Indonesia and officially known as the Moluccas, soon became a matter of national importance. Particularly so for Portugal, whose naval dominance, its rulers believed, should be mirrored by a similar commercial primacy.

While the trading of spices can be traced back to antiquity, it was Portuguese explorer Vasco da Gama who, between 1497 and 1499, first found a viable sea route from Europe to India, via the Cape of Good Hope. This easterly route remained the sole passage for the next 20 years, until an ambitious minor aristocrat, Ferdinand Magellan, entered the fray. Magellan set out on a voyage that, in finding a westward route, would change the way Renaissance

Europe saw the rest of the world and, crucially, complete the first circumnavigation of the Earth.

Many subsequent circumnavigations by ship have been motivated by commerce of one form or another; a ship, after all, is designed to carry things, be it goods or passengers. In 1693, the Italian Giovanni Francesco Gemelli Careri became the first tourist to pay for passage around the world. The recreational precedent set by Careri later became one of the principal forms of circumnavigation. Today, global cruises are a multi-billion-dollar business, bearing holidaymakers aboard luxurious floating cities; Royal Caribbean's *Symphony of the Seas*, in 2019 the world's largest liner, boasts a 228,081 tonnage that is nearly twice the industry average, and has capacity for 9000 high-spending passengers. The costs are more than financial, however, with destinations from the Bahamas to the Arctic Circle reporting environmental damage caused by the immense visiting vessels.

Others have sailed – the term applies loosely – the world compelled not by the views from the oceans' surface, but by what lies beneath; in 1960, at the height of the Cold War, the submarine USS *Triton* became the first vessel to circle the globe underwater. The tale of that tense voyage, and a variety of others here featuring mutinous crews, quicksilver trading and champagne jellies, seem to bear out the words of author Joseph Conrad, who served in the merchant navy: 'There is nothing more enticing, disenchanting, and enslaving than the life at sea.'

# FERDINAND MAGELLAN AND THE RENAISSANCE SPICE RACE

What does an ambitious Portuguese sailor do when his once glittering career hits the doldrums? Gamble everything at the rival Spanish court, promising the king a new route west to riches untold, and set sail with his violent, treacherous crew on what was to become the first-ever voyage around the globe.

Ferdinand Magellan was born around 1480. Home was the town of Sabrosa, in northern Portugal, where his father, Pedro, was mayor. In 1492, at the age of 12, Magellan, along with his brother Diogo, were appointed pages to the Lisbon court of Queen Leonora, consort to King John II. It was, of course, a symbolic year. The court was abuzz with stories, of Portugal's great rivalry with neighbouring Spain, and news of a voyage under the command of a certain Christopher Columbus, who had set off to discover new lands. Three years later, aged 15, Magellan entered the service of John's successor, Manuel I, whose accession owed as much to luck as to lineage.

## EARLY CAREER

By the time Magellan was planning a pioneering voyage of his own, he had become an experienced seaman. In 1505, once more in the company of his brother, the Magellans sailed to India, in support of Francisco de Almeida, the viceroy of Portuguese India who established Portuguese dominance in the Indian Ocean. For the best part of a decade, Magellan roamed the region, turning up across southern India, including the strongholds of Goa and Cochin, in Kerala. In 1506, he fought, and was wounded, in the Battle of Cannanore, now the Keralan city of Kannur. Clearly unafraid of a scrap, he was wounded again three years later, in 1509, at the Battle of Diu, in what is now Gujarat. The injury left Magellan with a permanent limp. (→ p.16)

Opposite: Portuguese (and later Spanish) explorer Ferdinand Magellan was at the forefront of the Spice Race.

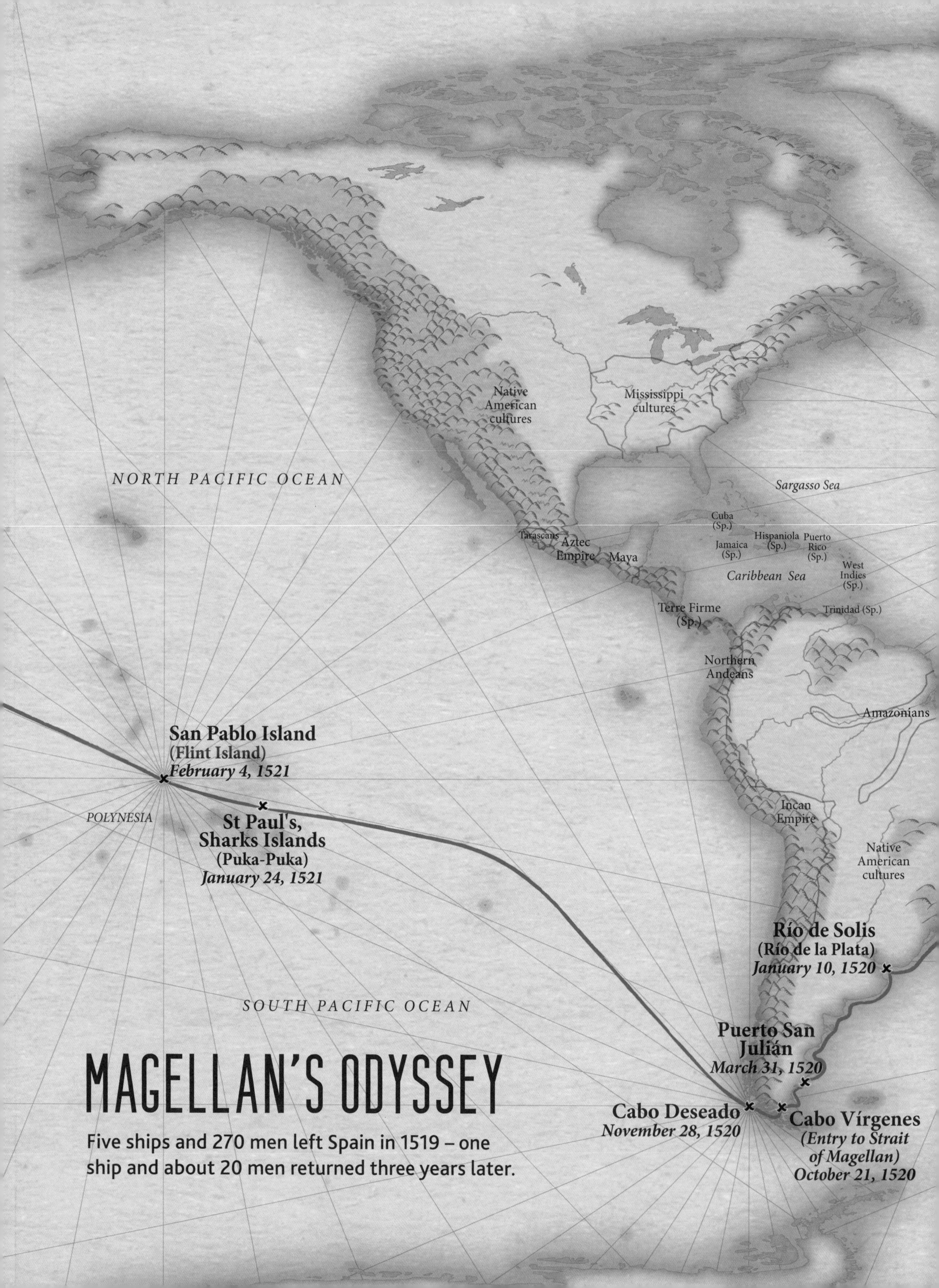

NORTH PACIFIC OCEAN

Native
American
cultures

Mississippi
cultures

*Sargasso Sea*

Tarascans
Aztec
Empire Maya

Cuba
(Sp.)
Jamaica
(Sp.)
Hispaniola
(Sp.)
Puerto
Rico
(Sp.)

*Caribbean Sea*

West
Indies
(Sp.)

Terre Firme
(Sp.)

Trinidad (Sp.)

Northern
Andeans

Amazonians

**San Pablo Island**
**(Flint Island)**
*February 4, 1521* ✗

*POLYNESIA*

**St Paul's,**
**Sharks Islands**
**(Puka-Puka)**
*January 24, 1521* ✗

Incan
Empire

Native
American
cultures

**Río de Solis**
**(Río de la Plata)**
*January 10, 1520* ✗

*SOUTH PACIFIC OCEAN*

# MAGELLAN'S ODYSSEY

Five ships and 270 men left Spain in 1519 – one
ship and about 20 men returned three years later.

**Puerto San**
**Julián**
*March 31, 1520* ✗

**Cabo Deseado**
*November 28, 1520* ✗

**Cabo Vírgenes**
**(Entry to Strait**
**of Magellan)**
*October 21, 1520*

*Greenland Sea*

*Barents Sea*

*Ka...*

Greenland

*Norwegian Sea*

Saami

Iceland
(Nor.)

Denmark-
Norway

Sweden

Livonian Order

Muscovy

Sibir

Scotland

Kazan

Ireland England
Wales

Holy Roman
Empire

Poland and
Lithuania

Astrakhan

Kazakhs

France

Venice
Papal
States

Hungary

*Black Sea*

*Caspian
Sea*

Uzbeks

Portugal **Spain**

Sardinia Naples
Sicily

Georgia

Ottoman Empire

Afghans

**Sanlúcar de Barrameda**
*Depart September 20, 1519*
*Arrive September 6, 1522*

**Seville**
*Depart August 10, 1519*
*Arrive September 8, 1522*

Cyprus

Safavid Empire

Multan
Su

*ORTH
LANTIC
CEAN*

Morocco

Sulaym

Muscat
(Port.)
Oman

Samma

Rajputana

*Islas Canarias
(Sp.)*

**Canary Islands**
*September 26, 1519*

Berbers

Arabs

Gujarat

M.
Kh

Ahmadnagar

Bija

*Sahara Desert*

Hawwara

*Arabian Sea*

Goa (Port.)

Arguin
(Port.)

Cabo Verde
(Port.)

Sennar Abyssinia Tahirids

Cannanore (Port.)
Calicut (Port.)
Cochim (Port.)

Great
Fulo
Gorée
(Port.)

Songhai

**lha de Santiago,
Cabo Verde**
*July 9, 1522*

Mali

Borgu

Bornu

Kwararafa
Oyo Igala

Shewa
Damot Ifat

Warsangali

Colombo (
Matara

Cacheu
(Port.)

Bonoman
Akan
Gold Coast
(Port.)

Benin
Nri

Bamum
Mandara

Principe
(Port.)

Fernando Pó
(Port.)

African
cultures

Ajuran

São Tomé (Port.)
Annobón (Port.)

Bunyoro Africa

ndaré
ort.)

Buganda

Burundi

Malindi (Port.)
Mombasa (Port.)

São
Salvador
(Port.)

Nroyo
Kongo

Kilwa

St Helena

Maravi

Quelimane (Port.)

**Santa Lucia Bay**
(Rio de Janeiro Bay)
*December 13, 1519*

Mutapa
Butua

Sofala

Madagascar

**Cape of Good Hope** ✕
*May 19, 1522*

*SOUTH ATLANTIC OCEAN*

*SOUTHERN OCEAN*

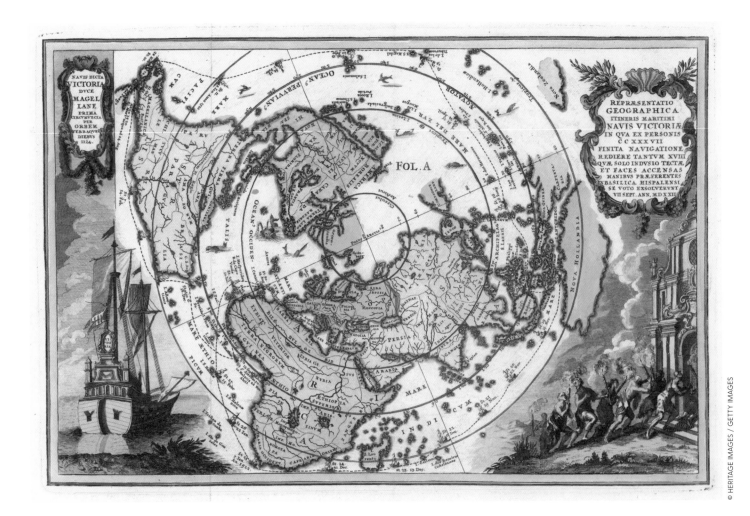

By 1512, he had ranged as far as the Malay Archipelago, where he fought in the capture of Malacca (not to be confused with the Moluccas) under Afonso de Albuquerque, the powerful Portuguese governor. It was in Malacca that Magellan 'acquired', and subsequently baptised, a local man, known to history as Enrique of Malacca. Engaged as Magellan's manservant for the subsequent decade, Enrique would go on to occupy a significant footnote in maritime history.

A key player of the period was Magellan's friend – and according to some sources, his cousin – Francisco Serrão. In 1512, Serrão was part of a Portuguese expedition sent by governor de Albuquerque to find the 'new' spice islands of Banda, 1930km (1200 miles) east of Java, then the only known source of mace and nutmeg. Serrão settled, married a local woman and became a military advisor, which is to say mercenary, to the Sultan of Ternate, a city state in the Moluccas. Serrao's letters home to Magellan, proclaiming the Moluccas' abundant spice harvests, would prove crucial in securing state patronage.

The following year, 1513, Magellan was among the 15,000-strong force that sailed for Morocco under the flag of Manuel I. The 500-strong fleet soon overwhelmed forces loyal to the Moroccan

## TALL STORIES

In Patagonia, Magellan and his crew, according to expedition chronicler Antonio Pigafetta, came across a tribe of 10ft-tall 'giants'. They kidnapped two to return to Europe, but the 'giants' died aboard ship. It's now believed that Magellan had encountered members of the Tehuelche people, of statuesque size, but typically in the 6ft range.

governor, who had refused to pay a yearly tribute to the Portuguese empire. Magellan's choice to stay on in the country would propel him, literally and figuratively, in a new direction.

## FALL AND RISE

Having served his king loyally, Magellan suddenly fell out of favour with Manuel's administration. Trumped up charges appeared; specifically, that Magellan was operating an illegal network with local Arab traders. He was eventually cleared but also found himself unemployed. For all his seafaring experience, Magellan found nobody willing to engage his services, at least not in a position suited to a man of his standing.

Despite fractious relations with Manuel's court, Magellan proposed an expedition. He contended, rightly, that by sailing in a westerly direction, a faster route might be found to the spice islands, one that avoided the arduous southern tip of Africa. There were, initially, grounds for optimism. This, after all, was the Age of Exploration and Manuel had form as an enthusiastic and ambitious patron. He had been instrumental in da Gama's voyage to India; his sponsorship had brought success for another explorer, Pedro Alvares Cabral, the first European to set foot in Brazil, in 1500. And ongoing Portuguese dominance in the Indian subcontinent was the result of Manuel's continued favour.

Yet Magellan's petitioning fell on deaf ears. In 1517, after repeated pleas, he severed relations with Manuel for good. He turned instead to Portugal's great rival, King Charles V of Spain. This was a teenage monarch, aged just 17, but one who also happened to be the Holy Roman Emperor. (And who famously claimed, as a rising star in the House of Habsburg, to speak Spanish to God, Italian to women, French to men and German to his horse.)

Magellan was an operator, of that there is little doubt. Upon his arrival in Seville, in October 1517, he knew nobody of any note and, unlike the king, spoke no foreign languages. Ever the pragmatist, he engineered a friendship with a compatriot, Diogo Barbosa, and soon married Barbosa's daughter, Beatriz. Barbosa was an intimate of a Spanish court vigorously pursuing a policy of expansionism, including ambitions to dilute Portuguese influence in the Spice Islands.

Through Barbosa, Magellan engineered an audience with Charles V. Armed with the letters from Serrão, he would have been all too aware he was speaking to the grandson of King Ferdinand and Queen Isabella, the

Spanish monarchs who, in 1492, had bestowed such generous patronage on Columbus. Charles agreed to fund a westward voyage; in return, Magellan, who renounced his Portuguese nationality in favour of a Spanish one, dangled before his benefactor the prospect of untold riches. With funding secured, Magellan set about poring over sea charts, enlisting the help of a cosmographer, Rui Faleiro. Together, they hoped to identify a channel through the South American continent, linking the Atlantic to whatever lay beyond; it may well have been the Age of Exploration but maps and charts of the period still decorated unknown parts of the planet with dragons and sea monsters.

The 'Armada de Molucca' struck out from the port of Sanlúcar de Barrameda, in southern Spain, on 20 September 1519. The fleet ran to five vessels, the *Trinidad*, the lead ship under Magellan's personal command, escorted by the *San Antonio*, the *Concepción*, the *Victoria* and the *Santiago*. Of the 270 men aboard, only 40 were Portuguese. It would prove a dangerous imbalance when sailing into the unknown. For that was an unequivocal fact: as Magellan set sail, the Earth had only two known oceans, the Atlantic and the Indian.

## MUTINY AT EASTER

Sailing west, they arrived in Rio de Janeiro in December. Three months of searching followed, Magellan's ships hugging the coast as they sought a sea route through the South American land mass. The onset of winter forced the fleet to shelter in Puerto San Julián. The five months they spent there were, according to expedition chronicler Antonio Pigafetta, full of argument. The captains of the other vessels, Spaniards all, frequently contested Magellan's authority. How Spanish the former Portuguese might have felt when his pragmatism came back to haunt him, Pigafetta does not record. At midnight, on Easter Day 1520, a vicious mutiny broke out, led by Juan de Cartagena, captain of the *San Antonio*, and abetted by Gaspar de Quesada, in charge of the *Concepción*. Alliances were hastily forged, then just as quickly betrayed; sailors hopped between factions, on occasion luring recruits before murdering them; at one point, only two of the five-ship fleet remained loyal to Magellan.

When he finally regained control, Magellan's sentencing was severe and, as ever, a little self-serving; de Quesada was beheaded but de Cartagena, whose uncle was a powerful archbishop, Juan Rodriguez de Fonseca, was merely marooned. Albeit in remotest Patagonia, where he was never heard from again. (Nearly six decades later, upon his own arrival in Puerto San Julián, Francis Drake claimed to

Clockwise from top left: the southwestern coast of Isla Grande, Chilean Tierra del Fuego; Magellan's circumnavigation mapped in Heinrich Scherer's 1702 *Atlas Novus*: spices such as nutmeg were premium products during the 15th century; windblown flag tree, Patagonia; Beagle Channel lighthouse, Ushuaia Bay.

have discovered the gallows used by Magellan to execute those, Drake assumed, who were lower-ranking mutineers.)

A less single-minded man might have begun to think events were conspiring against him: having put down the mutiny, Magellan sent the *Santiago* ahead to explore the coast, only for it to sink in rough seas; and when he finally found a route through the South American continent, in what is now southern Chile, the severe winter weather caused the crew of the *San Antonio*, under new but no less treacherous command, to desert and sail back to Spain.

Above: the baptism of Cebu dignitary Rajah Humabon. Below: the Strait of Magellan. Opposite: Magellan's fleet 'discovers' the Pacific Ocean.

## SAILING OFF THE MAP

Today, the Strait of Magellan's vital statistics are known to all mariners; the principal route is 560km (350 miles) long and between 3km and 32km (2-miles and 20-miles) wide, at its narrowest and widest points respectively. Navigating their way through the numerous, uncharted channels, it took Magellan's remaining crews fully one month to plot a course. Their reward, in November 1520, was a great body of water Magellan named, on account of its balmy conditions, Mar Pacifico, the Pacific Ocean. On reaching the Pacific, Magellan had literally sailed off the map. He had done so the moment he entered the strait which would be named in his honour. Magellan's guess was that this new ocean might be crossed in as little as three days. In the event, the Pacific leg of Magellan's voyage took nearly four months. As many as 30 men died from scurvy, although Magellan himself, in part thanks to privileges available to him as captain, was also able to call on the large quantities of quince fruits he preserved for the voyage.

## THE MISSIONARY

On 6 March 1521, Magellan's three remaining ships reached the island of Guam. Exhaustion amplified an already restive mood and altercations with the native Chamorro soon took place. The Spaniards were, to put it mildly, forthright in response, resulting in a number of Chamorro fatalities. Having resupplied,

Magellan sailed on to the Philippines, making land on the island of Cebu. Here, he made friends of the local dignitaries, including the imposing Rajah Humabon.

If Magellan's westward voyage was an imperial mission, the inseparable nature of church and state meant it was also a Catholic one. Surrendering to a periodic bout of religious fervour, Magellan enlisted the help of his new acquaintances to convert the indigenous population to Christianity. It must have been a fervour indeed for, as Pigafetta tells it, Magellan was able to convert more than 2000 souls in just a few weeks.

According to Catholic Church records, Spanish missionaries first brought Christianity to the area in the 1560s. The mass Magellan held on the island of Limasawa, on 31 March 1521, suggests otherwise. News of the event would almost certainly have found its way back to Charles V and Pope Adrian VI; even today, the population of the Philippines is predominantly Roman Catholic. (Modern-day Cebu City, in the Philippines, is twinned with Sabrosa, Magellan's home town.)

Pigafetta, for his part, recorded a parallel version of events, one that included beach orgies involving Magellan's crew and local women, and written up with lusty enthusiasm.

## PLIGHT OF THE NAVIGATOR

Not everybody shared Magellan's new-found zeal. The nearby island of Mactan, under their fearsome chief, Lapu Lapu, was particularly hostile. When the Cebu people asked for Magellan's help in combating their bellicose rival, he immediately agreed. Buoyed by the skirmishes in Guam, and against the advice of his men, Magellan now styled himself not as a pioneer of exploration but a man of war. Putting faith not just in his god but in more technologically advanced European weapons, such as muskets and mortars, he assumed victory would be swift and decisive.

On the morning of 27 April 1521, Ferdinand Magellan sailed to Mactan with a small flotilla. A rocky approach to the beach forced them to wade ashore, where the 49 men who hauled out were met by at least 1500 angry Mactanese. Magellan's men retreated but not before their leader was struck by a poisoned spear. As he sank face down into water, the Mactanese warriors rushed in to finish him off.

## THE SPICE ISLANDS

With their leader slain, the two remaining ships readied themselves for the onward voyage to the Moluccas. An expedition plagued by internecine plotting and betrayal was to suffer

©FABIO LAMANNA / SHUTTERSTOCK

yet another twist; unhappy at Magellan's failure to subdue Lapu Lapu, Rajah Humabon thought it perfectly reasonable to poison the remaining crew. Choosing the farewell feast as the venue for his ambush, Humabon managed to bump off up to 30 men, including many of what still passed for the expedition's officer class. Among those who died was Duarte Barbosa, son of Diogo, the courtier who had arranged Magellan's audience with the Spanish king. Some historians suggest Humabon was aided in his skullduggery by Magellan's former manservant, Enrique of Malacca, whose demands for his freedom following his master's death were denied, in direct contravention of Magellan's wishes.

## THE SPICE REACHES SPAIN

The two remaining ships, the *Trinidad* and the *Victoria*, reached Tidore, in the Moluccas, in early November 1521. (The *Concepción*, woefully short of crew, had been scuttled in the Philippines.) Laden with heavy cargoes, they set sail for Spain a month later. Almost immediately, the *Trinidad* began to take on water and both ships returned to port. An inspection of the *Trinidad* revealed serious structural damage and the decision was made that the *Victoria*,

## DRAKE'S PROGRESS

Thanks to Magellan's death in the Philippines, the first man entirely to sail around the world as captain of his fleet would be Francis Drake, six decades later. Ordered by Queen Elizabeth to harry Spanish forces on the Pacific coast of the Americas, Drake embarked from Plymouth in 1577, returning in September 1580 with a vast fortune.

under the command of Juan Sebastián Elcano, should return to Spain alone. The vessel arrived back in Seville a year later, in September 1522. Her vast haul of spices, as much as 26 tonnes, remained intact but only a handful of the 270 crew who had departed three years previously, around 20, were still aboard. Despite the trials it had endured, and inflicted, the expedition was, financially, a great success. Even allowing for the loss of four vessels, the project was deemed to have made a profit of more than 500 gold ducats – the equivalent spending power today of nearly $63m.

## SO, WHO WAS FIRST?

Given Magellan did not complete the voyage, credit for the first circumnavigation of the Earth must surely go to Juan Sebastián Elcano, the man who finally piloted the *Victoria* up the Guadalquivir River into Sanlúcar de Barrameda. There are, however, several technicalities. A reasonable case is also to be made for Enrique of Malacca. 'Purchased' by Magellan a decade before the Portuguese-cum-Spaniard's death, Enrique had sailed from Malacca to Spain in his master's employ. He had subsequently made the voyage across the Atlantic, through the Strait of Magellan and then crossed the Pacific.

© ARCHIVE PHOTOS / GETTY IMAGES

Some sources even suggest Enrique was not from Malacca at all but rather from the Philippines, where he had been captured, as a boy, by slave traders. Were this true, it would mean that, in returning to the Philippines once more, in 1521, the first true circumnavigation of the world belonged to him, albeit a somewhat truncated one.

## THE AFTERMATH

Magellan's body was never recovered. Despite entreaties from Pigafetta and others, Lapu Lapu proved resistant to the last, refusing to release either Magellan's corpse or his personal effects. One reading of Pigafetta's journals suggests that the Mactan chief planned to keep the body as a war trophy. Perhaps one of the great figures of Renaissance exploration suffered the indignity of being displayed in a corner of Lapu Lapu's home. The chieftain, in his resistance of Spanish colonial ambitions, went on to enjoy folk-hero status.

To Pigafetta, it seemed 'that every evidence of Ferdinand Magellan's existence had disappeared from the Earth'. It is certainly true that Magellan was not long mourned by any family back home. His two sons, Rodrigo and Carlos, both died in infancy while Beatriz, who was born the year Columbus set sail, passed away in 1522, the year her husband's expedition returned.

Opposite: Indonesia's Banda Islands, in the Maluku (formerly Molucca) Archipelago. Below: Ferdinand Magellan meets his end at the hands of the Mactanese.

Ferdinand Magellan himself might have taken some comfort from his legacy. For one thing, there are the expedition's remarkable achievements, which today are doled out as so much historical fodder in high schools the world over. Which is to do them a disservice.

In making a complete circumnavigation of the world, Elcano, Enrique or whoever else might lay claim, helped realise the full extent of the Earth. Their records put the total distance of the voyage at 14,460 Spanish leagues, or 60,450km (37,560 miles). Given we now know that the circumference of the Earth at the Equator is 40,075km (24,901 miles), it shows that, for all the in-fighting and intrigue, somebody was paying attention. It is a detail supported by the expedition's other significant discovery – that the Earth has differing time zones. On its return to Spain, the *Victoria's* log, which had been diligently maintained, showed quite clearly that the expedition had lost an entire day. This was due, of course, to the fact they were travelling west, in the opposite direction to the Earth's rotation. Such was the furore that met this discovery, a special delegation was hastily convened to explain the phenomenon to a bemused, and highly sceptical, Pope Adrian.

The scientific outcome of this revelation was, ultimately, the establishment of the International Date Line, the line of longitude at roughly 180 degrees, east of which is one day earlier than west. In passing through the Strait of Magellan, the expedition also discovered, from a European perspective, the extent of the Pacific Ocean. Thus they connected the Eastern and Western hemispheres, stitching the Old World to the New.

## THE LEGACY OF MAGELLAN

The Spanish and Portuguese reached the Spice Islands nearly a century before the arrival of the next great wave of imperial powers, the Dutch and the English. When Spain promptly turned her avaricious attentions toward South America, it meant that, ironically for Magellan, it was Portugal that reaped the greatest immediate benefit of the westward route.

That a post-colonial assessment treats Magellan less favourably is not surprising. The nature of his death was no less than he deserved, so the argument goes. But there is respect, too, if grudging, for the man himself. 'Many of those deaths and the loss of several ships might have been avoided had Magellan lived,' wrote the historian W Jeffrey Bolster. 'Only a megalomaniac of Magellan's stature could have set the enterprise in motion... only a mariner of his technical skill, intuition and icy leadership could have kept the fleet seaworthy in the daunting conditions it faced.'

# HE DIED 500 YEARS AGO. BUT FERDINAND

## THE MAGELLAN STRAIT

The 560km (350-mile) Magellan Strait runs through what is now Chile and has a maximum depth of around 1000m (3280ft). It is 3.2km (2 miles) wide at its narrowest point, and 32km (20 miles) at its widest extent. An important route for international shipping it continues to be a source of political tension, both with neighbouring Argentina and in turn between Britain and Argentina, who contest its use due to the Strait's close proximity to the Falkland Islands.

## MAGELLANIC PENGUIN

Magellan's men had to learn through trial and error that this strange 'black goose' required skinning, not plucking, before serving it at the captain's table. Few recipes, or reviews, survive, so we assume the penguins eaten by Magellan's men tasted like chicken. Fishy chicken. Magellan was the first European to observe the Magellanic penguin, in 1520. The birds themselves dine on cuttlefish and krill and are the largest example of the genus *Spheniscus*, which includes Galápagos and African jackass penguins.

## THE MAGALLANES

The Navegantes del Magallanes, known simply as 'the Magallanes', are one of the original franchises in the Venezuelan Professional Baseball League (LVBP). Founded in 1917, just four years after the New York Highlanders were rebranded to become the New York Yankees, the Magallanes is the oldest sports club in Venezuela, playing out of the 15,000-seater Estadio José Bernardo Pérez, in Valencia.

## THE MAGELLAN RAILCAR

Official rail transport of US presidents between 1943 and 1958, the Ferdinand Magellan Railcar (aka United States Railcar 1) was first used by Franklin D Roosevelt. The first presidential railcar since that commissioned by Abraham Lincoln in 1865, Magellan was armour-plated and fitted with bulletproof glass. Adapted from a special-edition Pullman carriage, one of six named after celebrated explorers, the rear platform was used by Harry S Truman to make speeches during political campaigns. It was decommissioned after Eisenhower's presidency.

## THE MAGELLAN PROBE

The first deep-space probe launched by NASA's Space Shuttle programme, in this case *Atlantis*, the Magellan probe's mission was to carry out close planetary reconnaissance of the surface of Venus. By the time it had completed the third of its three fly-bys, between 15 May 1991 and 13 September the following year, the Magellan probe had mapped an extraordinary 98% of the surface, far exceeding NASA scientists' hopes for the mission.

# MAGELLAN IS EVERYWHERE...

## THE LAND OF MAGELLANICA

Until Captain Cook reached Australia – or rather, Terra Australis – in 1770, the existence of a southerly landmass was largely hypothetical. Its presence was first postulated as far back as the 5th century; geographers believed there to be a strange 'south land' as a necessary counterweight to the northern hemisphere. It began to appear regularly on maps during the Renaissance; of the various names it carried down the centuries, one of the more enduring was 'Magellanica', land of Magellan.

## THE MAGELHAENS CRATERS

Magellan has given his name to three impact craters in our solar system, two on the surface of the Moon and the third on Mars. The International Astronomical Union (IAU), which oversees celestial nomenclature, insists on Greek or Roman for land features. Craters however, take the name of the person they are named for; the Moon is home to Magelhaens, identified in 1935, and Magelhaens A (2006), while the Martian surface has a Magelhaens (1976) all its own.

## THE MAGELLAN BIRDWING

The Magellan Birdwing is a large butterfly native to the Philippines. First recorded by the artist and zoologist Robert Henry Fernando Rippon in Icones Ornithopterorum, his early 20th century illustrated guide to the butterflies of the region, the Magellan Birdwing is one of a number of species whose wings display a spectacular iridescence when viewed from an oblique angle.

## THE MAGELLANIC CLOUDS

The Magellanic Clouds is the name given to two dwarf galaxies that have likely been observed by humans for millennia. They were named for Ferdinand Magellan around 1800 and are part of the southern celestial hemisphere. Both are in orbit around the Milky Way, with the greater system, the Large Magellanic Cloud, estimated to be 160,000 light years away from Earth. Estimates put the lesser system, the Small Magellanic Cloud, at around 200,000 light years away.

# TICKET TO RIDE

Frustrated by the slow progress of his career, Giovanni Careri embarked on a 17th-century sabbatical to make the first recreational circumnavigation of the world: he hit paydirt in the Philippines, penetrated secretive China and was wowed by the ancient ruins of South America.

The man credited with the first recreational circumnavigation of the planet by ship might, in another life, have been a travelling salesman. In the late 17th century, an Italian named Giovanni Francesco Gemelli Careri paid his passage around the globe on a series of cargo ships. He is occasionally touted as a potential inspiration (one of many) for Jules Verne's *Around the World in Eighty Days*.

Not many details are recorded about his life, although we know that he trained to be a lawyer. Born in Calabria in 1651, Careri studied at a Jesuit college in Naples but was hindered from rising through the legal ranks by, as he saw it, insufficiently aristocratic lineage. So, in 1693, he took the only other option available to men of means – he decided to take a gap year. Or rather, a five-year sabbatical.

His plan for this grandest of tours was to purchase goods in a given destination, then sell them on for a profit at his next port of call, where they might be in demand. From Persia, for example, Careri took dates to India, for a modest profit; from the Philippines, he transported a quantity of quicksilver to Mexico, for which he saw a return of 300% on his investment.

Careri was an observant and enquiring traveller who, by his own admission, was more than a little fortuitous. Having crossed the Middle East and India, Careri went next to China, where Jesuit priests mistook him for a Vatican spy. Thinking they might aid some papal mischief, the Jesuits parlayed Careri's entry into 17th-century China's closed and secretive society. He visited the Great Wall, which he found inferior to Rome's Colosseum, before arriving in Macau, from where he sailed for Manila.

His good fortune continued. On reaching Central America and buoyed by the handsome rewards of his quicksilver trading, Careri inveigled his way into the circle of Don Carlos de Sigüenza y Góngora. In New Spain, in Mexico City, Sigüenza was a leading intellectual who, in his capacity as royal geographer to the Spanish court, had produced the first map of the territory; Sigüenza was also a champion of indigenous culture and thus indulged Careri with a tour of the ancient ruins of Teotihuacán.

The Italian eventually made his way home and Careri's account of his trip, *Giro del Mondo*, was published in 1699. It ran to five volumes, one for each year of his journey. He died in Naples in 1725.

Above: Giovanni Francesco Gemelli Careri, as depicted in his 1699 book *Giro del Mondo*. Below: illustration of an Aztec calendar from the 1791 French translation, *Voyage du Tour du Monde*. Opposite: Cunard's *Britannia* sails out of Boston in 1847.

# THE FIRST GLOBAL CRUISE: FOR 'MILLIONAIRES' ONLY

Cunard supplied the luxury, American Express provided the investment, and 450 passengers paid lavishly for champagne jellies and oyster patties for the 1922 voyage via the Panama Canal, Far East, Suez Canal, Mediterranean and Atlantic (first class only).

Cunard, a byword for luxury travel, started life as a postal service. Established in 1839 by Samuel Cunard in partnership with celebrated Scottish shipbuilder Robert Napier, it was awarded the first British transatlantic mail contract (as the British and North American Royal Mail Steam-Packet Company) the following year. The *Britannia*, a paddle-steamer that first set sail from Liverpool on 4 July 1840, was the first scheduled shipping service across the Atlantic. The line's four paddle-steamers operated on the Liverpool–Halifax–Boston route (Cunard himself was from Nova Scotia) and enjoyed something of a monopoly. Offering a 35-guinea fare, the fleet held various Blue Ribands for the fastest Atlantic voyage for the best part of three decades. Charles Dickens was famously a passenger, sailing on *Britannia* in 1842.

In the 1870s, competition arrived in the shape of the White Star Line, when that company switched its focus from voyages to the Australian goldfields to the transatlantic arena. In 1879, in a bid to raise the capital required to regain supremacy, Cunard was relaunched as the Cunard Steamship Company. The early years of the 20th century saw a power struggle to control commercial shipping, with mergers and

© TOPICAL PRESS AGENCY / GETTY IMAGES

© TOPICAL PRESS AGENCY / GETTY IMAGES

counter-mergers on both sides of the Atlantic; names tempted into the fray included Lehman and JP Morgan.

The money lay in passengers and goods; specifically, getting as many of both across the Pond as fast as possible. Before WWI, undertaking voyages for pleasure was far from a mass-market activity. There were well-established leisure classes on both sides of the Atlantic; the Vanderbilts were staggeringly rich by the 1820s, with the Rockefellers close behind. Yet it remained a bespoke affair. Those wishing to see the Rococo wonders of Versailles, or the temples of India and Southeast Asia, were required to send cables and charter private vessels. (So-called 'round-the-world cruises' originating in the US often involved lengthy, cross-continental rail travel.)

The best part of a century after it waved off the first scheduled mail service across the Atlantic, the Cunard Line inaugurated another pioneering voyage. In early 1922, the American Express Company approached Cunard with a request. Like Cunard, American Express had started life as a mail company, following the merger of two rather well-known courier companies, those owned by Henry Wells and William G Fargo. Diversifying into luxury travel (by

## FINAL VOYAGE

The *Laconia* was torpedoed in 1942 off west Africa with 2725 aboard: its crew, British military personnel, civilians, Polish guards and Italian prisoners of war. The U-boat captain responsible rescued 400, before a US aircraft attacked, killing many survivors. In a major turning point of WWII, U-boats were subsequently ordered not to pick up survivors.

1903, its $28 million of assets were second only to those of the National City Bank of New York), American Express asked to charter the *Laconia*, a 2200-berth passenger-liner built to sail between New York and Southampton, to where Cunard had relocated to better serve the affluent London market. *Laconia* was small by modern standards, at around 182m (600ft) long, with a 20,000 tonnage. (Cunard's current flagship, the *Queen Mary 2*, has a 150,000 tonnage).

*Laconia* set sail on 21 November 1922. Cunard made much of the ship's route, following in the wash of Ferdinand Magellan voyage; these were not just luxury tourists but pioneers. Typically, the busiest decks of Atlantic liners were steerage, those occupied by third-class passengers. For Cunard's world cruise, the emphasis was on luxury. There would be no third-class travel – only first. Through their agent, Thomas Cook, American Express billed the trip as a floating private members' club, to which membership would be limited to just 450 paying passengers.

The press portrayed it as a cruise for millionaires; in fact, the number didn't much exceed Cunard's standard first-class capacity, of around 350 passengers. Most aboard the *Laconia* were middle-class Americans.

© E.F. CORCORAN / GETTY IMAGES

Opposite: clay-pigeon shoot from the decks of the *Laconia*.

Clockwise from left: the *Laconia's* luxurious salon; steaming out of Liverpool; Merchant Navy Captain Hossack on the *Laconia's* bridge.

*Laconia* was one of four Cunard ships to undertake world cruises in 1922 and 1923. And the vessel did secure genuine firsts during its voyage, albeit less consequential ones; *Laconia* was, for example, the first liner of its size to negotiate the Panama Canal. From there it headed across the Pacific (there were no stops in Australia or New Zealand) on the way to the Far East, Suez, the Mediterranean and the Atlantic. Passengers dined on first-class staples, such as oyster patties, pressed beef and galantine of game, washed down with champagne jellies.

The *Laconia* returned to the United States on 30 March 1923, more than four months after her departure and having stopped at 22 ports. Since that inaugural world cruise, the Cunard Line has carried more passengers on global cruises than any other company.

The *Laconia's* part in that record, however, was limited. On 12 September 1942, now converted to a wartime troopship, she was torpedoed by a German U-boat and sank. The submarine's commander, Werner Hartenstein, promptly set about attempting to rescue the 2725 passengers aboard but an estimated 1600 people lost their lives (see sidebar on opposite page for more).

# PROPELLED BY HISTORY

From transatlantic speed records to naval deployments, Cunard has been synonymous with international cruising for 180 years.

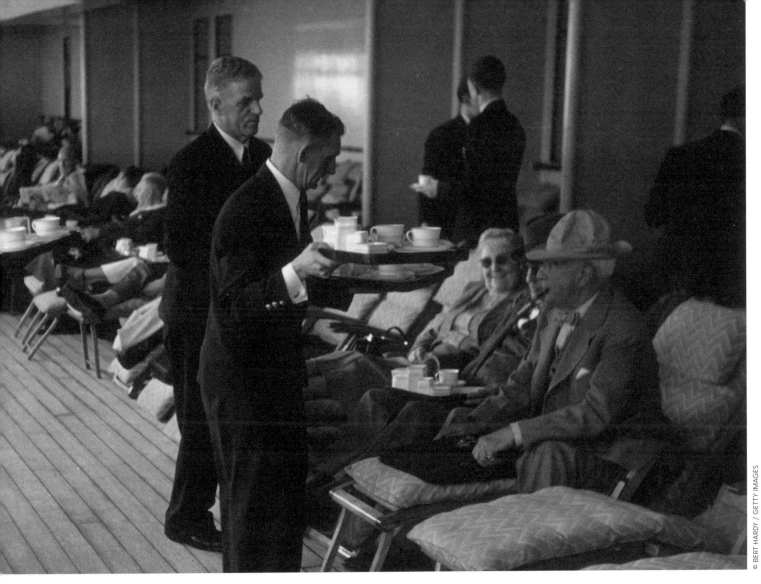

Opposite: posing under the propellers of Cunard's brand-new *Queen Elizabeth II*, 1967.

Clockwise from top: taking tea on the deck of Cunard's *Queen Elizabeth*, 1948; the *Franconia's* Edwardian gym, 1911; dockside goodbyes as the *QE2* sails for the Falklands, 1982.

# FROM HARD TIMES TO LEISURE TIME

On Charles Dickens' first translatlantic trip in 1842, everything was rough: the crossing, the mutton and the company. By contrast, today's 200,000-tonne cruisers are vast pleasure palaces catering for every desire.

When Charles Dickens sailed from Liverpool to Boston aboard the *Britannia*, in October 1842, he found the first scheduled steamer service across the Atlantic far from the luxurious experience of modern-day liners. Drinking took place around the clock, as did the card schools, although the rough weather they encountered – fully two weeks of it – meant potential winning hands often fell to the floor. Much like Dickens' fellow passengers and, as Dickens reported, the *Britannia*'s drunken crew; while the more expensive fares included all alcoholic drinks, even steerage passengers paid little for their pints of stout.

Dickens was withering about the service, reporting cabins that leaked so much seawater he had to store his possessions in the sink, and a pillow like 'a muffin beaten flat'. (It is possible his opinions were coloured by the subsequent experience of his first US tour, which turned into an unqualified disaster.) The menu, Dickens noted, ran to little more than boiled mutton, mouldy apples and 'plates of pig's face'. He was grateful for fresh milk, however, provided by a cow that made the voyage suspended in a hammock.

The modern liner, by comparison, is a veritable floating city. Its premium offering is a voyage not simply across one ocean but around the world, aboard vessels so well provisioned as to make your regular landlubber's life look positively drab. Royal Caribbean's *Symphony of the Seas*, as of 2019 the world's largest liner, boasts 2759 cabins and can carry close to 9000 passengers and crew. It is the length of four soccer pitches, features 18 decks and 23 swimming pools. In fact, *Symphony* has seven distinct 'neighbourhoods', designated variously for sports and leisure, retail, dining, entertainment and so on. And where Dickens might note, ruefully, that pretty much everything is available around the clock.

The *Symphony*'s 228,081 tonnage is nearly twice the industry average – it came at a build price of $1 billion – and dwarfs the *Queen Mary 2*, Cunard's flagship launched in 2004 and for a time the world's largest around-the-world liner. In 2019, with a paltry 150,000 tonnage, the *QM2* failed to make even the top 10. The *QM2*'s staterooms, however, would likely please Dickens, offering as they do the choice of 10 types of pillow.

*See overleaf for how to build a cruise ship.*

## 01
### On the upside
A cruise ship is constructed on land, in sections weighing up to 800 tonnes each. It is put together upside down, for the simple reason of gravity – it's easier for shipbuilders to weld heavy steel plates downward rather than up. The cabins, restaurants and other internal architecture are also built on shore, then manoeuvred into what becomes a floating shell. Despite the fact cruise ships can be built, in some cases, in as little as six months, demand has for many years outstripped supply. An estimated 27m people took a cruise in 2018 and with industry growth of 20% over the five years to 2019, cruise operators simply can't build vessels fast enough, at least until the coronavirus crisis of 2020.

## 02
### Does my bow look big in this?
A cruise ship is a sailing billboard for its operating line. This means it needs to look distinctive and yet, seen on the horizon, most cruise ships look the same. One way of standing out is to give the ship an identifiable silhouette. The liner *Celebrity Edge*, launched in 2018, has a vast bow that tilts forward, giving her the appearance of a giant hatchback car. The *Quantum of the Seas* meanwhile, has a 300ft-high observation 'pod'. Her sister ship, *Symphony*, is pretty unmistakeable by way of her sheer size.

## 03
### Plastic fantastic
Each disaster brings new implications for industry safety. Combustible materials have long been a no-no. Fixtures and fittings are either plastic, light metal or ceramic. Other precautions include sealable bulkheads and numerous sets of fire doors. Other threats include piracy, particularly off the Horn of Africa, although most ships currently avoid the region, or else turn off the lights.

## 04
### The non-negotiable
Nobody goes on a cruise to swim inside. Pools are always located on the uppermost deck, both for sunbathing and sea views. The downside is weight, at 28kg per cubic foot, and located a long way from the ship's centre of gravity (*Symphony* has 23 pools). Despite this, top-deck pools remain the first names on the team-sheet when it comes to cruise-ship design.

## 05
### Old ladies of the sea
Designed for 20 years' service, and sometimes as many as 30, a ship's premium status tends to be diluted as she ages, not least because newer vessels come on to the market. American passengers prefer their ships relatively gimmick-free and thus they tend to have a classic silhouette that evokes a traditional, which is to say, European heritage. European passengers, on the other hand, seem to readily embrace quirkier designs.

# ANATOMY OF A CRUISE SHIP
Building ocean-going ships is big business: these are some of the facts and stats behind the behemoths.

### Facts and figures

Big ships need a big push. *Symphony* is powered by four 14,400kw diesel engines, and an additional pair of 19,200kw engines. The ship's 2200 crew must navigate galleys and storerooms, which are squeezed into just 20% of the ship's internal space. The 40 onboard restaurants get through a staggering 9000kg of potatoes a week. (Presumably they serve other dishes.)

### Captain's bird's eye

When he's not entertaining guests at the captain's table – still a big thing, apparently – the captain can be found in his quarters. These are located near the bridge area, so the captain is always available, particularly in an emergency or in heavy weather. Often near the top of the ship, it affords him birds'-eye views but is also the part of the vessel that rolls most in bad weather. Increasingly larger vessels mean this is not the problem it once was.

### Green cruising

Cruising can hardly be considered eco-friendly but some companies are slowly adopting more sustainable goals. Most encouragingly, Hurtigruten has commissioned its new expedition ships with full hybrid power capability to reduce emissions and nine of its ships are being retrofitted with hybrid engines. A few Carnival cruise ships can be plugged into shore-side electrical power when in port so they don't need to keep the engine running. And companies are also investing in exhaust scrubbing filters and banning single-use plastics.

### Rolling is rare

Weight distribution is key, particularly on modern liners, whose multi-deck design would otherwise make them susceptible to capsizing in rough seas. Weight pulls a ship down, while buoyancy pushes it up: too much weight and it sinks; too much buoyancy and it topples over. The centre of gravity therefore is very low, a counterweight of engines, fuel and stores. The higher the deck, the lighter the materials. Cinemas, for example, are made up of plastic seats and fresh air. The two most well-known examples of cruise ships rolling, the *Costa Concordia* in 2012, and the SS *Eastland*, which sank after rolling in port in Chicago in 1915, are separated by almost 100 years.

# WHY YOU MIGHT BE BETTER OFF SAILING INTO THE SUNSET

Retirees have long been fond of a leisurely cruise. But with the cost of urban living ever increasing, young and old alike are being targeted with year-long, all-in, around-the-world deals aboard floating cities.

A cruise around the world used to be a once-in-a-lifetime experience. Today, sailing continually around the globe can be cheaper than renting an apartment in some of the world's major cities. That's why some passengers, notably elderly retirees, spend more time cruising than they don't. Beatrice Muller famously lived on the *Queen Elizabeth 2* for eight consecutive years, from 2000 to 2008, when the ship was retired from service.

According to a 2016 report in Britain's *Daily Telegraph*, average outgoings for people living in central London, including rent, utilities, food, shopping and leisure, were around £97.71 ($128) a day. A full-board, round-the-world cruise, including a single-person's supplement and lasting 120 days, the industry standard, came in at £88.53 ($116) per day. Planet Cruise, meanwhile, an online specialist one can assume has a vested interest, published a report, in 2017, that went a step further, comparing the costs of various round-the-world itineraries with those of conventional city life. For an entire year.

If you're prepared to effectively move house once a month, the savings seem impressive. One option, aboard 12 individual four-week cruises, took in 163 destinations in 35 countries on five continents. The costs for the year came in at $31,280, including full board and entertainment, such as regular cinema visits and an annual gym membership.

Compared with the expense of life in 10 of the world's major cities, Planet Cruise estimated passengers could save as much as $48,000 a year by living at sea. That, however, was compared to living in Monaco, shorthand for exorbitant pricing, and where basic yearly outgoings per person were, according to the report, in the region of $130,000.

London residents would save a more modest £5700 ($7458) during a year afloat, while New Yorkers would see both the waves and their bank balances swell by around $17,000. A year's cruising was only $865 cheaper than living in Abu Dhabi for the equivalent period, and the high-spending citizens of Los Angeles – presumably not the legions of out-of-work actors – found themselves only $1790 better off.

Veteran cruisers have long perfected the routine of prepaying bills, stocking up on medication and getting their mail held for months at a time. And round-the-world cruises are no longer entirely populated with retirees, according to industry advocates. With internet access turning ships into floating offices, cruises attract an increasing number of entrepreneurs, although how that works should you need to arrange a face-to-face meeting at short notice is unclear.

© MARCO SECCHI / GETTY IMAGES

# TROUBLED WATERS

**If mass tourism represents an invading army, then cruise liners are the troopships. So what is the industry doing to clean up the enviromental havoc it is wreaking?**

Cruises ships bring tourists. And then some. In the most well-publicised cases, the damage wrought by the disproportionate numbers of visitors to local residents has raised questions about the viability of the industry. Take Venice, Italy, a destination which has long been the canary in the coalmine. It is estimated that, since the 1980s, around half the population of the historic old city has left. Back then, resident Venetians numbered around 100,000. They have been replaced by approximately the same number of visiting tourists, and on a daily basis. In high season, around one third of those, in excess of 30,000 people, will disembark from cruise ships, driving prices up and remaining residents mad. The problem is not restricted to historic European port-cities, nor is it simply one of scale.

Operators are keen to exploit the growing market for 'wilderness' cruises. The Galápagos have long been a popular destination, as have the polar regions. Key operator Hapag-Lloyd promotes its Arctic fleet as 'small ships', although each takes around 200 passengers into otherwise pristine environments. Campaigners point to knock-on effects. In 2016, the liner *Crystal Serenity*, with capacity for 1000 passengers, was the first cruise ship to sail through the fragile Northwest Passage.

In terms of environmental impact, the cruise industry is almost always in deep water. Major liners hold 3000 passengers as a matter of course; many can berth double that, and some three times as many. The resources required simply to feed and bathe them are vast, the biggest vessels producing 1200 tons of waste every day. Environmentalists claim around 1 billion gallons of waste is dumped into the sea every year.

In December 2015, the *Zenith* liner, carrying 1800 passengers, demolished significant coral reefs off the coast of Grand Cayman when it dropped anchor there. As before, the issue is not just about numbers. In 2018, there was outrage when a polar bear was shot dead after it injured a guide who had been escorting cruise passengers on a land excursion in Norway's Arctic Archipelago.

Is a solution to be found? In Venice, it was announced in 2018 that large ships will be required to moor further out to sea, though this seems little more than a sticking plaster. Also in 2018, the Hurtigruten line launched MS *Roald Amundsen*, the first of two hybrid-powered liners and a crucial step for the eco-cruising market. Advertised as the safest, greenest expedition ships ever, as of 2019 it was still promoting voyages through the environmentally sensitive Northwest Passage – the first hybrid vessels to do so – as well as to Antarctica. In a global industry worth around $126 billion, there seems little incentive for wholesale change.

Above: during high season, some 30,000 cruise-ship passengers crowd into Venice each day.

# HOW THE GREAT WHITE FLEET REINVENTED GUNBOAT DIPLOMACY

**It began as a way to cool sailors below decks on USS Dolphin in the tropics. The white hulls caught on – within 20 years they were conspicuous around the world, a symbol of President Theodore Roosevelt's desire to assert American might on the high seas.**

The Great White Fleet was an armada dispatched not for battle but diplomacy; or at least, a diplomatic mission that anticipated future conflict. The brainchild of Theodore Roosevelt, a president with one eye on his legacy, the plan was to give a rapidly emerging US Navy practice in circumnavigation, communication and fuel-management. If it sounded like a beginner's course in deployment, it was.

In the years after the American Civil War, the US Navy had become something of a joke in harbours and embassies around the world. By 1880, noted the naval historian Frank Bennett, 'repairs were no longer possible, for space for more patches was lacking upon almost every ship afloat.'

The US government's response was to build four ships, the first of which, the USS *Dolphin*, was sent to the tropics, in 1888. It proved instructive. When the *Dolphin*'s crew began to struggle with the heat inside a black, steel hull, the captain broke with naval protocol and repainted the ship – with white lead. The effect was to cool the interior by several degrees. Soon, all new warships were making their maiden voyages in white, in what became the US Navy's official peacetime colour.

By the turn of the 20th century, US naval vessels were still primarily intended for coastal defence. In 1886, maritime historian Captain Alfred Thayer Mahan was appointed head of the US' newly established Naval War College, in Newport, Rhode Island. Mahan argued that the chief business of a navy at war was to seek out potential enemies and destroy their fleets.

Accomplishing this meant long-range deployment and that meant big ships. In 1906, the British Royal Navy introduced a ship that changed the rules of engagement: HMS *Dreadnought*, extremely fast thanks to steam-driven turbine engines, and extremely well-armed, equipped exclusively with armour-piercing 12in guns. The template for all ships that followed, it rendered everything before it obsolete.

For all Mahan's bellicosity – America's first dreadnought, USS *Michigan* was already underway and launched in 1908 – Roosevelt knew history showed something else; that power could be exercised softly, while carrying a big stick.

Naval courtesy calls were just that, honouring a host monarch or head of state with an entire visiting flotilla, whose task was primarily pageantry. It was also an iron fist in a velvet glove, the gentle rattling of sabres. There were precedents for Roosevelt's mission. In 1891, the large French fleet that called on Russia's Tsar Nicholas II, in Kronstadt, did so against a backdrop of 30 years of hostilities between the two nations. The tsar was so impressed, however, he signed a treaty with the French three years later.

The US joined the party in 1902, when Roosevelt invited the German navy to send a squadron to New York City. A reciprocal invitation promptly followed, for US ships to participate in similar fleet 'celebrations', not just in Germany but in Britain and France.

Long before Pearl Harbor, Roosevelt had begun to look west. In 1905, Japan had announced itself as a major sea power by defeating, at Tsushima, a Russian fleet many naval experts believed unbeatable. America had strategic interests in both the Philippines and wider Pacific theatre. Roosevelt wanted it known that a powerful American fleet could be deployed at any time, able to defend US interests from both its Atlantic and Pacific ports.

It was all theoretical, of course, the US having never attempted a naval operation on this scale. It was true that US maritime prowess had been bolstered by success in the Spanish-American War of 1898, when under Commodore George Dewey the US had defeated Spain's Pacific Fleet in the Philippines. But a greater global deployment brought different challenges. Not only would it involve the operational capability of the entire US Navy but there was no guarantee the different classes of vessel could even sail in formation.

Another pressing issue was fuel. Unlike the European imperial powers, the US Atlantic Fleet couldn't call upon a global network of coal stores in its colonies. The solution proved diplomatically fraught. In shipping 125,000 tons of coal to San Francisco, Roosevelt was forced to call largely on British supply ships. Britain, hedging its bets, had responded to Japanese success at Tsushima by

Above: Roosevelt's Great White Fleet dreadnoughts take to the seas. Below: the French navy visits Russia's Tsar Nicholas II at Kronstadt, 1891.

signing an alliance with Japan that same year.

Roosevelt's 16-strong, white-clad fleet set sail from Hampton Roads, Virginia, on 16 December 1907, bound for the British West Indies, then on to Brazil and Chile. The Panama Canal was still seven years from completion, forcing the fleet to plot a course through the Strait of Magellan. They followed the coast north, as far as San Francisco, making for an impressive sight, in particular the 14,000 sailors who paraded on deck whenever the fleet made port. They left San Francisco in July 1908, bound for Honolulu, followed by New Zealand and Australia.

Thousands of well-wishers turned out to see the spectacle. The US fleet visited three cities in Australia (the country's own navy was granted royal approval by George V in 1911). Roosevelt claimed major political coups, too. In Messina, in Sicily, crews from four of the presidential fleet aided the recovery effort following the devastating 1908 earthquake. Among the many bodies recovered was that of Arthur Cheney, US consul to Italy.

By far the greatest diplomatic success came in October that year, when the fleet docked in Tokyo. The Japanese proved surprisingly welcoming, although their own fleet, daubed in dark war paint, greatly unsettled Roosevelt's officers. (Within months of the fleet's return, in February 1909, the entire US Navy was repainted in the colour we now call battleship grey.)

Its white predecessors, meanwhile, had covered some 79,636km (43,000 nautical miles) on their voyage around the world, calling in 20 ports on six continents. Moreover, they demonstrated the US was becoming a major sea power. In 1908, 39 more dreadnought-class ships glided down US slipways; briefly, America's navy was second only in size to that of Britain's.

Roosevelt had proved it possible to talk softly, while sailing a big ship.

# UNDER PRESSURE: WHEN THE COLD WAR PLUMBED THE DEPTHS

**Russia, it was thought, was about to deploy its first nuclear missile-capable submarine. So, in 1960, the US Navy quickly threw the USS Triton into the underwater arms race on a secret mission to recreate Magellan's voyage underwater.**

The first circumnavigation of the planet underwater began on 24 February 1960. Its start – and subsequent end – point was St Peter and Paul Rock, an outcrop of land in the middle of the Atlantic. The vessel in question had left London a week previously but now Operation Sandblast, the name given to the mission by the US Navy, began in earnest. The voyage was part of a sub-aquatic space race, the vessel in question a watery Sputnik. In 1960, the US was in no doubt that the launch of Russia's first nuclear-missile capable submarine was imminent.

Their response was the USS *Triton*, a state-of-the-art vessel and one of only a handful to be powered by two reactors (all other twin-reactor subs were Russian). That made her fast, capable of 28 knots, but it was the $109-million price tag, not including nuclear reactors, that set Congressional pulses racing. The US Navy regarded speed to be key to the mission. The Polaris missile programme, by then fairly advanced, would require men to be submerged for long periods. With psychological stress as likely as physical trauma, they aimed to circle the globe in 56 days.

The mission took place in febrile political times: the Cold War was particularly frosty; the Paris Summit of 1960 was looming. With President Eisenhower due to meet his Russian counterpart, Nikita Khrushchev – who two years later, along with John F Kennedy, would bring the world to the brink of nuclear war during the Cuban missile crisis – Ike needed something to bring to the table.

*Triton*'s captain, Edward Beach, preferred a light touch. A scholar of history, he was a leader in the mould of Ernest Shackleton, not quite one of the boys but no dominant overseer either. The mission was hit by problems almost immediately. On the approach to Cape Horn, the senior radar operator, John Poole, began suffering from severe kidney stones. Next, *Triton* lost her fathometer, a crucial machine which monitored depth by echo-location of the seabed.

They passed Cape Horn on 1 March and struck out across the Pacific. The plan was to follow the route of Ferdinand Magellan. *Triton* next saw land as she passed Easter Island, where, through the periscope, the crew spotted the statue erected by Thor Heyerdahl.

Radarman Poole was eventually picked up by a passing US naval vessel and *Triton* crossed the International Date Line on 23 March. When they reached Guam, one crew member asked if he might use the periscope. Petty Officer Edward Carbullido had been born in Agat Bay, where the *Triton* lay submerged, and from beneath the waves he now observed his parents' house. It was the first time he had done so in more than a decade.

High-tech glitches were accompanied by mundane chores. Even a $100-million-dollar sub needs to take out the garbage; *Triton* fired hers from a torpedo tube. Fresh air came by way of a 'snorkel'. Where modern subs use generators to extract oxygen from seawater, the *Triton* was forced, so as not to be detected, to raise a special submarine snorkel at night in order to refresh the vessel's internal atmosphere.

Beach sailed next for the Philippines and, on 1 April, located Mactan Island, where Ferdinand Magellan had met his end at the hands of the Mactanese chief, Lapu Lapu (see page 19). Later that afternoon, on (perhaps appropriately) April Fool's Day, *Triton* had its one and only encounter with a sailor who did not have official clearance. Rufino Baring was a 19-year-old Filipino fisherman who was, that afternoon in the Pacific, simply paddling his own canoe. Quite what he thought when the *Triton*'s periscope broke the surface, just metres from where he sat with his lines in the water, is anybody's guess.

*Triton* headed next across the Indian Ocean. It was here they conducted one of their less conventional experiments. Beach had determined to cross the ocean without surfacing, not even for the air snorkel. To keep the air as fresh as possible, a brief smoking ban was introduced. The ship's doctor, Benjamin Weybrew, took the opportunity to carry out a series of stress tests; he aimed to establish whether or not, under trying circumstances, non-smokers coped better than smokers

The voyage finally ended on 25 April 1960, at the same rocky outcrop where it began. However, Eisenhower's planned celebrations had to be curtailed in the light of Russia's downing, a week later, of the U2 spy plane.

© BETTMANN / GETTY IMAGES

**Clockwise from top: Captain Edward Beach at the *Triton*'s periscope; celebrations after crossing the equator in February 1960; manoeuvres off Delaware Beach following completion of Operation Sandblast.**

*Triton*'s success brought forward the deployment of Polaris-equipped nuclear submarines, in 1960. Beach was awarded the Legion of Merit and the Magellanic Premium, the US' most prestigious award for advances in science. He also wrote *Around the World Submerged*, an account of his trip. In a footnote, three years after the voyage's end, Dr Benjamin Weybrew was invited to speak at the annual convention of the Cigar Institute of America. Asked how smokers fared under pressure when deprived of nicotine, he replied: 'Badly.'

© PJF MILITARY COLLECTION / ALAMY STOCK PHOTO

# BICYCLE

## BY

Circumnavigating the world by bike has, according to the endurance rider Mark Beaumont, long had a British flavour. The first man to complete a circuit of the globe on two wheels – Thomas Stevens in 1886 – originally hailed from Hertfordshire. Beaumont is a Scot and the ride he made, in 2017, took him round the planet in just 78 days, beating the time of his fictional inspiration, Phileas Fogg.

To suggest British domination of this mode of transport is misleading, though. Stevens, in fact, emigrated to the US as a child and grew up learning to ride penny-farthings on the unforgiving hills of San Francisco. The first woman to pedal round the globe, Annie Londonderry, also moved to America, from her native Latvia, at the age of four or five, setting off on her circumnavigation from Boston some 20 years later.

Where British cyclists have distinguished themselves, perhaps, is by displaying a certain eccentricity. There's Ben Page, for example, who in 2015 was one year into an unscripted if conventional circumnavigation when he secured sponsorship from a manufacturer of 'fat' bikes, whose outsize wheels enabled him to complete his dream by riding around the world almost entirely off-road. Of those who have kept to the road well-ridden, a little idiosyncrasy has taken them a very long way; one rider Mark Beaumont encountered in 2017 was Ed Pratt, a then 21-year-old making his way around the world by unicycle.

A factor likely to deter potential cycling circumnavigators is the sheer amount of kit required for months or years in the saddle. As a general guide (for an unsupported ride), it's a matter of laying out on the floor everything you think you will need before eliminating the many items you can't conceivably carry. Accommodation plans will amount to nights under canvas or in a ditch; on the menu, day in, day out, will be camp-stove cuisine.

For the sort of rider who does not tend to have the time to cycle the entire planet, organised events which allow them to accrue their circumnavigation one landmass at a time present a good alternative. Such continent-wide 'sportives' are well spread geographically, from the incomparably brutal Race Across America – the record for which stands at an eye-popping 4800km (3000 miles) in a little under eight days – to its more modest, 3200km (2000-mile) European counterpart, the Transcontinental.

These two races are united in their riding philosophy: head down; ride hard; sleep wherever. But, rider, be warned – a circumnavigation by bike, by its very nature, amplifies daily cycling hazards, such as commuter traffic, and this chapter includes tragic tales of fatalities among recreational riders and racers alike.

# ONE MAN. TWO WHEELS. 385KM PER DAY...

When Mark Beaumont proposed a cycling circumnavigation in less than 80 days, many thought the idea as fictional as Phileas Fogg himself. But the Scotsman was adamant: all he had to do was to create a support team that would allow him to concentrate on cycling an eye-watering distance, day in, day out, for two and a half months.

Surprisingly enough, the man who has cycled around the world faster than anybody else also has an obsession with being decidedly average. When, in 2017, Scotland's Mark Beaumont set out to beat the record for a global circumnavigation by bicycle, his stated aim was to complete the journey in less time than the 80 days of Jules Verne's fictional hero, Phileas Fogg. To do so, Beaumont reckoned, he simply had to maintain an average daily speed and distance. But meeting these averages would mean setting down a punishing plan that required him to ride the equivalent of New York to Boston every day – and to do so for two and a half months straight.

## RAP SHEET

As an adventurer, Beaumont had form. In fact, he had already made a circumnavigation by bicycle, in 2007–08. Back then, the record for a 29,000km (18,000-mile) circuit of the globe stood at 276 days. And although that amounted to riding your bike 105km or more (65 plus miles), day in, day out, for more than nine months, it was a target Beaumont thought eminently attainable. His effort saw him shave 82 days off the record, coming home in 194 days (and after riding an additional 480km, or 300 miles, to ensure he did not fall short). Contending with the usual inconveniences befalling the independent traveller – an upset stomach in Asia, the occasional theft – he would look back on that trip as a naive adventure, and ridden at a pace, 160km (100 miles) a day for six months, that was positively plodding.

Left: Mark Beaumont and his bicycle on The Mall in London before setting off on his round-the-world challenge.

It was also the ride that changed the course of his career. A successful, if modest, television career followed, as did other cycling expeditions; across North and South America and, in 2015, an attempt on the classic cycling route across Africa, from Cairo to Cape Town. Beaumont smashed the record (of course) carving an impressive 17 days off the time when he reached Cape Town in just 42 days.

Not content with wheels, Beaumont next turned his attention to water. In 2012, he joined the crew of Atlantic Odyssey, a six-strong team bidding to break the record for crossing the Atlantic Ocean. The crew had logged 27 of a planned 29 days at sea, and a distance of 3704km (2000 nautical miles), when their boat capsized in rough weather. But it was that first cycling circumnavigation that played on Beaumont's mind as he rowed and, in particular, the thought that he could do it more quickly. Much more quickly.

## HOW FAR AND HOW FAST?

Adhering to the old adage about eating an elephant (one bite at a time), Beaumont decided to break the required 29,000km (18,000 miles) down into four-hour sets, of which he would ride four every day. That meant 16

### BIKE FOR SALE

To raise money for charity, Mark Beaumont sold his round-the-world bike. The listing for his Koga Kimera Premium on the auction site eBay described it as 'used' and having 'covered 18,000 miles without fault'. Bids reached £10,000 ($13,000) by the end of the first day. It eventually sold for an undisclosed sum.

hours of daily riding, during which Beaumont calculated he could maintain an average speed of 24km/h (15mph). With the remaining hours each day given over to eating and sleeping, this meant an average distance of 385km (240 miles) every 24 hours. No mean feat, given that would be the equivalent, by way of a British reference point, of riding from London to Plymouth every day for ten weeks. Once those numbers were lodged in his head, however, Beaumont stuck to them slavishly.

By undertaking such a formidable challenge, Beaumont was risking more than a lack of sleep. At the time of planning, he was recently married, to Nicci, with whom he had a young child. In addition to the physical strain he would be placing himself under, the emotional trial was also significant; his commitment to the law of averages told him that spending so long on the open road increased the likelihood of riding into trouble.

By 2017, the record belonged to New Zealand's Andrew Nicholson, a former Olympic speed skater. Nicholson had completed his ride two years earlier, in 2015, unsupported and in a time of 123 days. Aware of the Kiwi's mark, as Beaumont plotted his course, he faced another predicament. If his ride went according to plan, he wouldn't simply beat Nicholson's mark but obliterate it. Fellow cyclists urged caution, suggesting Beaumont promote his ride differently, aiming for perhaps 85 or 90 days, so as not to appear arrogant. But Beaumont took the view that the 80-day mark was so iconic that failing to beat it would be regarded as an anticlimax.

## THE EXITS ARE LOCATED...NOWHERE

'I didn't want to have any excuses to fail,' he told Britain's *Guardian* newspaper, 'that's why I made the rule I would ride four blocks of four hours every day without any breaks. It makes life a lot simpler when you don't have any exit.' An important decision was to ride with a support team. The politics of ultra-endurance cycling are complex and divisive. More militant campaigners believe self-reliance is a key part of the challenge, and for that reason argue that only unsupported rides (see p.55) should be eligible.

For Beaumont, the only thing that mattered was speed. To that end, he assembled a veritable task force, specialists in logistics, bike mechanics and nutrition. A key member of the team was Laura Penhaul, former head physiotherapist for Britain's Paralympic team and with whom Beaumont had much in common. In 2016, Penhaul had been the leader of the Coxless Crew, the first all-female crew to row unsupported across the South Pacific, all 16,668km (9000 nautical miles). Penhaul's job description was simple: keep Beaumont riding. Responsibility for ensuring Beaumont's bike was ever-roadworthy fell to Alex Glasgow, a three-time British mountain-biking champion. Beaumont opted for a trio of bikes himself, all specially adapted with the help of Koga, the manufacturer.

Included on the permanent roster – part-time helpers on different stages brought the total number to 20 staff – was Mike Griffiths, a British army veteran whose experience in the Balkans, Falkland Islands and Afghanistan might prove useful if they encountered trouble; Griffiths had also served, more than once, as a logistics man for teams competing in the Race Across America (RAAM), the gruelling 4800km (3000-mile) sportive billed as the world's toughest bike race (see p.64).

## STAGE ONE: FROM RUSSIA WITH LUMPS

Beaumont set off from the Eiffel Tower in Paris on 1 July 2017.

Europe proved relatively uneventful, and he ticked off that part of his route, from Belgium to Latvia, in under a week. A seasoned endurance rider, Beaumont knew well enough that certain niggles, such as sore tendons or wrists (➜ p.50)

Below: Crossing North America. Opposite, clockwise from left: pedalling through Australia and Asia; checking routes around Moscow.

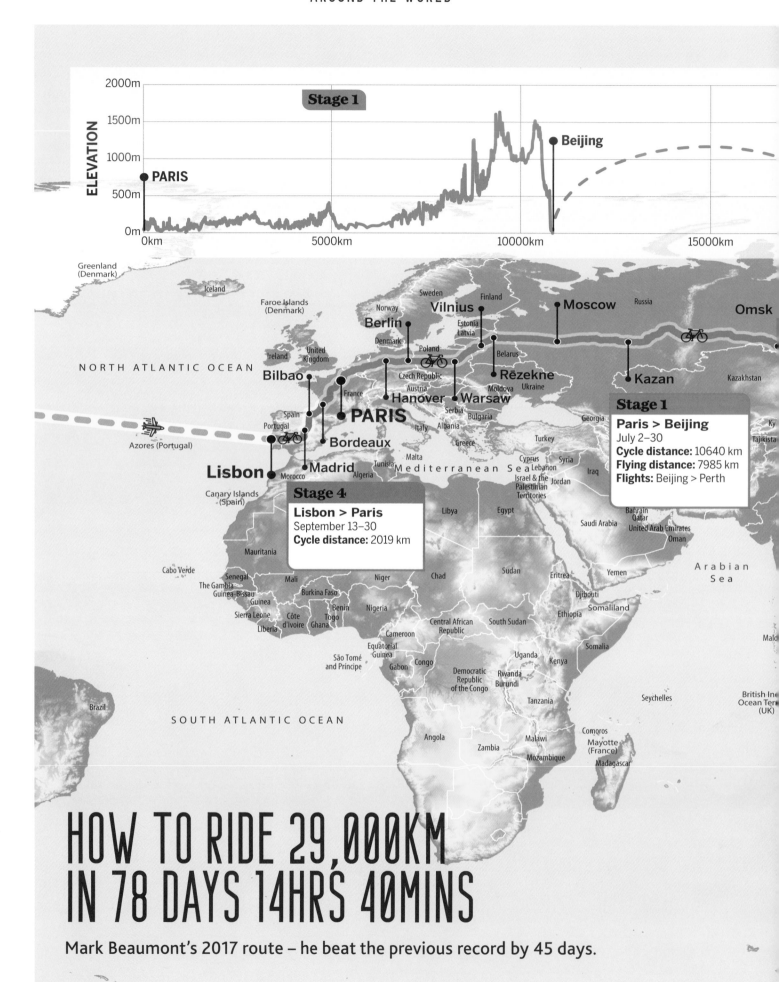

**ELEVATION**

Stage 1

PARIS

Beijing

0km   5000km   10000km   15000km

Greenland
(Denmark)

Iceland

Faroe Islands
(Denmark)

Sweden   Finland

Norway   Vilnius

Berlin   Estonia
Latvia

Denmark

Moscow   Russia

Omsk

NORTH ATLANTIC OCEAN

Ireland   United
Kingdom

Bilbao

France

Poland

Czech Republic

Belarus

Rēzekne

Kazan

Kazakhstan

Austria

Hanover   Warsaw

Spain

Portugal

PARIS

Serbia   Ukraine
Italy   Bulgaria
Albania

Moldova

Georgia

Ky

Tajikista

Azores (Portugal)

Bordeaux

Greece

Turkey

**Stage 1**

**Paris > Beijing**
July 2–30
**Cycle distance:** 10640 km
**Flying distance:** 7985 km
**Flights:** Beijing > Perth

Ky

Lisbon

Madrid

Morocco

Algeria

Tunisia   Malta

Mediterranean Sea

Cyprus   Syria
Lebanon   Iraq
Israel & the
Palestinian
Territories   Jordan

Bahrain
Qatar
United Arab Emirates
Oman

Canary Islands
(Spain)

**Stage 4**

**Lisbon > Paris**
September 13–30
**Cycle distance:** 2019 km

Libya   Egypt

Saudi Arabia

Yemen

Arabian
Sea

Cabo Verde

Mauritania

Senegal   Mali
The Gambia
Guinea-Bissau   Burkina Faso
Guinea
Sierra Leone   Côte   Benin
d'Ivoire   Togo
Liberia   Ghana

Niger   Chad

Nigeria

Sudan

Eritrea

Djibouti
Somaliland

Ethiopia

Mald

Cameroon

Central African
Republic   South Sudan

Somalia

Brazil

SOUTH ATLANTIC OCEAN

Equatorial
Guinea
São Tomé   Gabon
and Príncipe

Congo

Uganda
Democratic   Rwanda   Kenya
Republic   Burundi
of the Congo

Tanzania

Seychelles

British Ind
Ocean Ter
(UK)

Angola

Zambia

Malawi

Mozambique

Comoros
Mayotte
(France)

Madagascar

# HOW TO RIDE 29,000KM
# IN 78 DAYS 14HRS 40MINS

Mark Beaumont's 2017 route – he beat the previous record by 45 days.

# TEAM PLAYER

With a support crew on his 2017 circumnavigation, Mark Beaumont rode without luggage and was able to benefit from more complex and delicate technology.

Selle SMP Pro saddle with cutaway.

Koga Kimera Premium carbon fibre frame (size 60cm). Approx £3200 / $4200 complete.

Shimano hydraulic disc brakes for ease of maintenance.

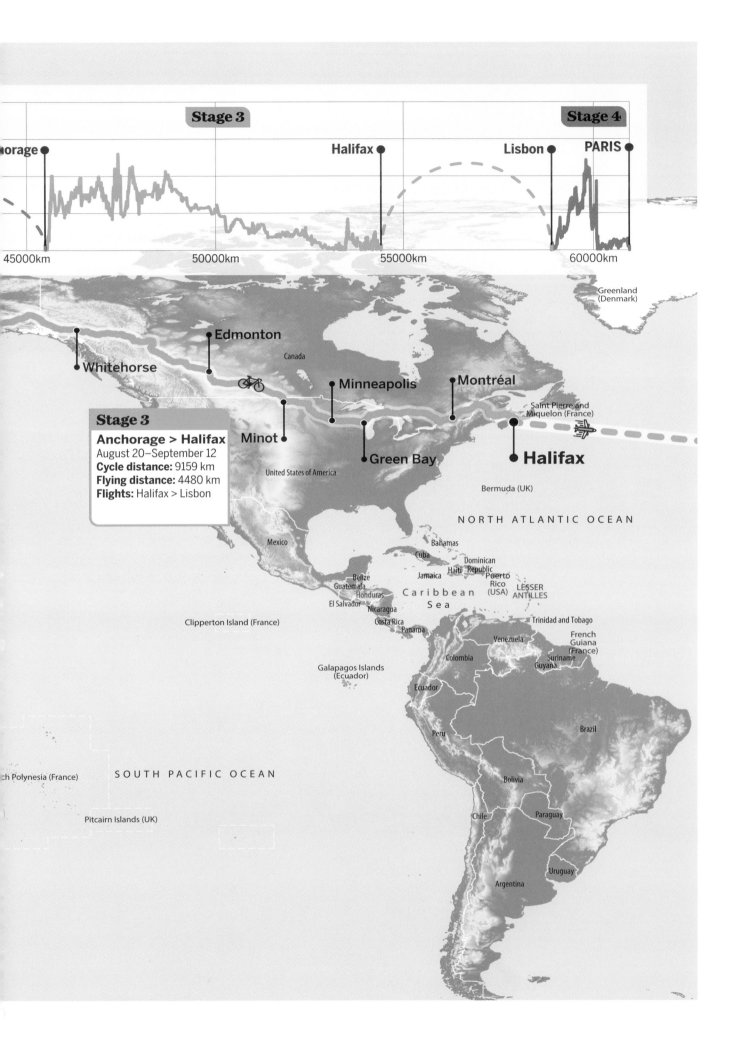

**Stage 3**

**Stage 4**

orage

Halifax

Lisbon

PARIS

45000km 50000km 55000km 60000km

Greenland (Denmark)

Edmonton

Whitehorse

Canada

Minneapolis

Montréal

Saint Pierre and Miquelon (France)

**Stage 3**

**Anchorage > Halifax**
August 20–September 12
**Cycle distance:** 9159 km
**Flying distance:** 4480 km
**Flights:** Halifax > Lisbon

Minot

Green Bay

United States of America

**Halifax**

Bermuda (UK)

NORTH ATLANTIC OCEAN

Mexico

Bahamas

Cuba

Dominican Republic

Belize
Guatemala
Honduras
El Salvador
Nicaragua

Jamaica

Haiti

Puerto Rico (USA)

LESSER ANTILLES

Caribbean Sea

Clipperton Island (France)

Costa Rica
Panama

Trinidad and Tobago

Venezuela

French Guiana (France)

Colombia

Suriname
Guyana

Galapagos Islands (Ecuador)

Ecuador

ch Polynesia (France)

SOUTH PACIFIC OCEAN

Peru

Brazil

Pitcairn Islands (UK)

Bolivia

Chile

Paraguay

Uruguay

Argentina

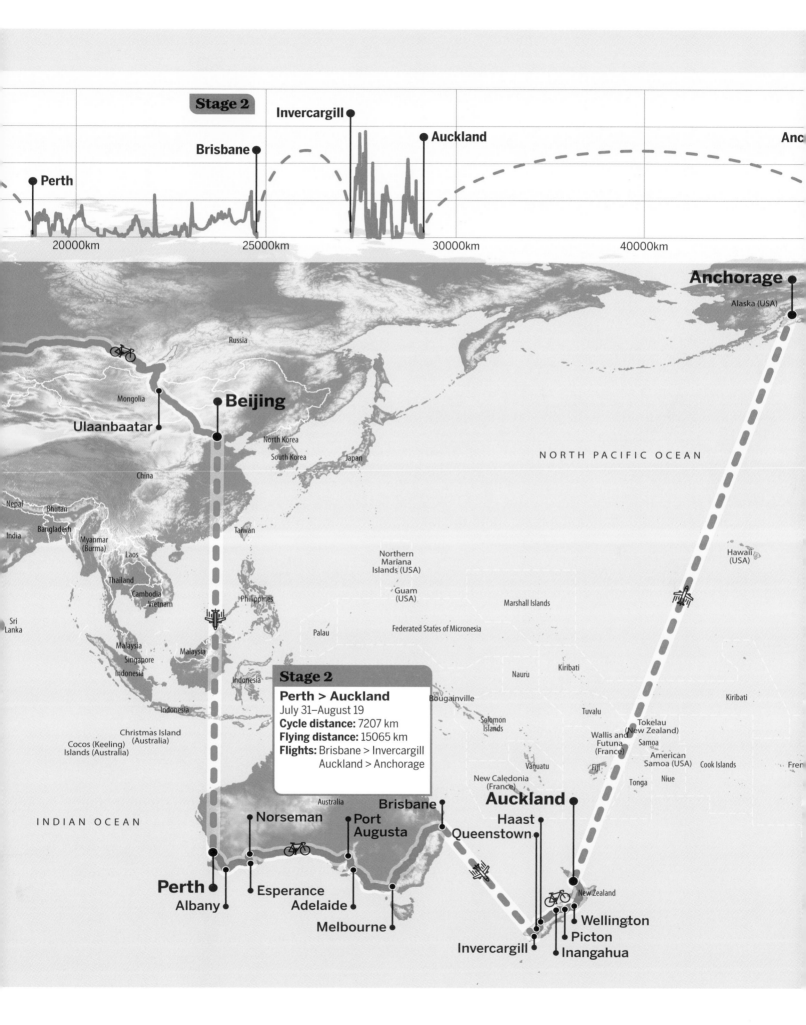

Stage 2

Invercargill

Brisbane

Auckland

Anc

Perth

20000km          25000km          30000km          40000km

Anchorage
Alaska (USA)

Russia

Mongolia

Beijing

Ulaanbaatar

China

North Korea

South Korea          Japan

NORTH PACIFIC OCEAN

Nepal

Bhutan

India          Bangladesh

Myanmar
(Burma)

Laos

Taiwan

Thailand

Cambodia
Vietnam

Sri
Lanka

Malaysia

Singapore          Malaysia

Indonesia

Philippines

Palau

Northern
Mariana
Islands (USA)

Guam
(USA)

Federated States of Micronesia

Marshall Islands

Hawaii
(USA)

Indonesia

Christmas Island
(Australia)

Cocos (Keeling)
Islands (Australia)

Nauru

Kiribati

Kiribati

Tuvalu

Tokelau
(New Zealand)

**Stage 2**

**Perth > Auckland**
July 31–August 19
**Cycle distance:** 7207 km
**Flying distance:** 15065 km
**Flights:** Brisbane > Invercargill
Auckland > Anchorage

Bougainville

Solomon
Islands

Wallis and
Futuna
(France)          Samoa

American
Samoa (USA)          Cook Islands          Fren

INDIAN OCEAN

Australia

Norseman

Port
Augusta

Brisbane

Vanuatu

New Caledonia
(France)

Fiji

Tonga          Niue

Haast

**Auckland**

Queenstown

**Perth**

Albany

Esperance

Adelaide

Melbourne

New Zealand

**Wellington**

**Picton**

Invercargill          **Inangahua**

Aero helmet from Rudy Project.

Aerodynamic handlebars by Profile Design.

28mm Panaracer tyres on Corima carbon rims for speed, light weight and comfort.

Shimano Ultegra 22-speed drivetrain with electronic shifting.

Aerodynamic handlebars for streamlined position.

Custom steel-framed Shand Stooshie bike frame, made in Scotland.

28-32mm Continental Grand Prix 4 Season tyres (six used).

# LONE RANGER

As a solo self-supported cyclist, Jenny Graham carried everything she needed on her record-breaking 2018 trip, selecting kit for its durability and ease of repair.

Weatherproof bikepacking bags from Apidura, Exped range.

Selle SMP Pro saddle.

Shimano Ultegra drivetrain (mechanical - two gear cables replaced).

Hydraulic disc brakes for all-weather stopping power.

caused by repetitive strain, would settle down within the first few weeks.

Not so the uncertain ones. By day nine, Beaumont was into Russia and was closing in on Moscow when he crashed heavily. The medical assessment showed that he had cracked a tooth and bruised his elbow. The ever-reliable Penhaul was on hand, patching him up and even replacing a filling that had come loose. An exhausted Beaumont managed to fall asleep during the impromptu dental procedure.

The rest of his trip through Russia did little to recommend the country as a riding destination. On routes riddled with potholes, his fellow road users showed scant regard for the well-being of a cyclist, even that of a circumnavigator, and his support team were frequently held up (by potholes, not bandits). Often riding into strong winds, Beaumont was in constant pain – subsequent MRI scans showed he had in fact fractured said elbow – and the adjustments he made to his riding position in order to alleviate the discomfort led, in turn, to nerve damage in his neck.

Beaumont was undeterred. 'My job was simple,' he told *Cyclist* magazine, 'get on the bike, ride for 16 hours, eat 9000 calories, sleep and repeat.' Simple indeed. Clearly this is a man wired differently to other athletes. However long the day ahead, he claimed that he only ever thought in terms of the next four-hour set, such was Mr Average's unwavering faith in the plan. With Russia behind him, it was time to step on it. Mongolia saw an upturn in Beaumont's spirits. He was joined periodically by herds of wild horses, galloping across a landscape dotted with yurts as he barrelled towards China. The Gobi Desert was a particular highlight.

Above: Mark Beaumont riding in a wintry South Island, New Zealand. Below: an impressive road-side companion in North America.

## STAGE TWO: AUSTRALIA AND NEW ZEALAND

Beaumont has always maintained his success was as much due to good planning as it was to mental and physical fortitude. In 2012, Guinness amended the rules governing around-the-world cycling records. Prior to that date, the regulations included the occasional restful hiatus, with the clock stopping during flights between continents. This allowed riders to catch up on much-needed sleep, without fretting about airport delays or cancellations. Under the new rules, the recorded time for a given attempt included flights, meaning a good riding schedule was nothing without a sound flight plan. Beaumont's reputation for attention to detail was well earned; at each point of disembarkation he employed fixers, local assistants designed to help him clear customs and get riding with the minimum of delay. In Australia, that meant he was up and riding within half an hour of touching down at Perth Airport.

Even with the best-laid plans, however, five hours of sleep will rarely offset 16 hours of riding. With tiredness a constant threat to Beaumont's record attempt, Penhaul drew on some tried and tested methods from her ocean-going adventures. 'Mouthwash seems to work,' she told the *Financial Times*, in an interview during the expedition, 'but Mark knows when that slow-blink starts. We'll get him in for a 10-minute catnap if he needs it.'

Despite a brief automotive hiccup, when the crew crashed the expedition's camper van, Beaumont traversed the southern coast of Australia without incident, from Western Australia to Brisbane, a route that took him through four of

# THE RECORD BREAKERS

## 1886

### Thomas Stevens
The first man around the world by bike was Thomas Stevens, who completed his ride aboard a penny-farthing between 1884 and 1886. Riding a Columbia 'Standard' whose front wheel measured 50 inches (127cm), Stevens had to walk as much as a third of the route.

## 1981

### Nick Sanders
The Briton set the record as we know it in 1981. But his 21,900km (13,609-mile) ride did not include the southern hemisphere and was completed in a time of 138 days.

© FAIRFAX MEDIA ARCHIVES / GETTY IMAGES

## 2012

### Mike Hall
In June 2012, after British endurance rider Mike Hall completed an unsupported global circumnavigation in 91 days, 18 hours, Guinness altered the rules to include total travel time, ie flights between continents, increasing Hall's time by over 16 hours. The attempt remains unratified by Guinness.

### Andrew Nicholson
A former Olympic speed skater, New Zealand's Andrew Nicholson notched up an unsupported bike circumnavigation in 2015, in a time of 123 days, 43 minutes.

© ROBERTO RICCIUTI / GETTY IMAGES

## 2012

### Juliana Buhring
Six months after Mike Hall's return, the British/German national Juliana Buhring finished her own unsupported circumnavigation in December 2012. Her time of 152 days, including flights, saw her become the first woman to complete a Guinness-ratified circumnavigation.

## 2015

## 2017

### Mark Beaumont
In 2017, Beaumont lowered Nicholson's mark by 45 days, coming home in 78 days, 14 hours and 40 minutes.

© ANDREW MATTHEWS / GETTY IMAGES

## 2017

### George Agate and John Whybrow (tandem)
Agate and Whybrow completed a global tandem ride in 290 days, 7 hours and 36 minutes in March 2017. Their record is under threat from Australians Lloyd Collier and Louis Snellgrove, though their 283-day equivalent ride in May 2019 is yet to be verified by Guinness.

## 2017

### Ed Pratt (unicycle)
Then 21, Somerset-born Ed Pratt survived the indignity of being overtaken by Mark Beaumont in the Australian outback in 2017. Pratt's epic, if ungainly, ride around the planet raised an impressive £300,000 for charity. His route covered 33,800km (21,000 miles) over 40 months.

### Jenny Graham
In 2018, Jenny Graham's 125-day ride broke the record for fastest woman to cycle unsupported around the world.

## 2018

the six states, a distance of more than 5000km (3100 miles). In the Outback he passed fellow Briton, Ed Pratt. The then 21-year-old was on the second of a successful three-year bid to become the first person to circumnavigate the globe on a unicycle. Yes, really. Beaumont pressed on, to a wintry New Zealand and some chilly hours working through his daily sets. Having cycled across both the North and South Islands, Beaumont flew to Anchorage, and the third leg of his journey, North America.

## STAGE THREE: ANCHORAGE TO HALIFAX

Rest periods, in the form of intercontinental air travel, brought their own problems. Beaumont's body was keen to heal when in fact he needed it to remain effectively under stress. During the ride across North America he began to suffer from 'hot spots', pressure points in both his hands and feet; at one stage in Canada, Penhaul was required to dig a callus from Beaumont's foot. In caring for her charge, Penhaul remained unflinching. She took daily swabs of Beaumont's saliva to continually monitor his health; and because the strain his body was under made him susceptible to infection, she sent advance instructions on personal hygiene to journalists who periodically joined Beaumont on his ride.

Just as they had been in Russia, the winds across the vast expanses of North America proved testing. Beaumont had plenty of time to do the maths. 'If a headwind drops my average speed by 5km (3 miles) per hour,' he told the *Financial Times*, 'I'm putting down more power to go slower. If you multiply that by 16 hours in a day, the losses can be huge. I can wake up and step out of the camper van at 4am knowing I'm going to spend the entire day... limiting damage.'

It was a grim time, Beaumont recalled later. Riding from Alberta to Saskatchewan, for example, an easterly wind reduced his planned 385km (240 miles) to just 190km. Desperate not to fall any further behind, Beaumont spent a gallant – or foolish – 17½ hours in the saddle. Strong winds continued for the next week.

As Beaumont pedalled his way across Canada, his route took him occasionally south, into the US states of North Dakota, Minnesota and Wisconsin. The scenery of the Great Lakes proved a balm although by now he was running on fumes. During his final two weeks in North America, Beaumont is convinced it was pure routine that got him through. 'I was in idiot mode,' he told *Cyclist*, 'devoid of emotion, neither excited nor depressed. That emotional response, the laughing, the crying, all happens in the first month. After that, you're left with nothing.'

Above, from top: there was to be no taking the train on Beaumont's feat; a massage table makes for an impromptu pitstop.

## STAGE FOUR: LISBON TO PARIS

It was doubtless a relief when Beaumont found himself aboard his fourth and final flight, from Halifax, in Nova Scotia, to the Portuguese capital of Lisbon. There was still the small matter of 1700km (1050 miles) of riding to be done in order to reach Paris, but the end was in sight.

## CLOSING ARGUMENTS

Beaumont returned to the Eiffel Tower triumphant, to be greeted by family and friends. He had ridden for 16 hours a day through 16 countries, logging a total distance of 29,030km (18,039 miles). The official time was 78 days, 14 hours and 40 minutes. (Or, as Beaumont would likely prefer, 471.5 four-hour blocks.)

His challenge thereafter was to adapt to a life not governed by numbers, to be able to simply ride his bike. And in this there was another plan to follow, a 'decompression' strategy that had been formulated by Penhaul – plenty of rest as well as reducing his daily calorie intake.

Like Phileas Fogg, Beaumont later returned to the Reform Club, a private members' club in central London, where he explained to a select audience how he had notched up his incredible feat. And as an addendum, the BBC informed Beaumont that the 11,315km (7031 miles) he had ridden between 2 and 31 July 2017 earned him the Guinness World Record for the most miles cycled in a calendar month.

Cycling around the globe had, Beaumont reflected, caused a strange disconnect in how he had come to see the world. The expanses he had covered were often vast in size; yet the fact he had done it so quickly made the world seem relatively small.

Beaumont cut an astonishing 116 days off his first record, yet he is adamant his new time is there to be broken. If a professional racing cyclist were to put his or her mind to it, Beaumont believes, his record would be in trouble. (That said, the longest ever stage of the Tour de France was 480km/300 miles, during the 1919 edition of the race; in 2019, the furthest the riders had to pedal during a single stage was a more modest 230km/144 miles). At the time of publication, Beaumont's record still stands.

While he'd beaten the fictional Fogg by almost two days, Beaumont admitted that his circumnavigation had taken him to some dark places. In two decades of endurance challenges – he'd cycled across Scotland at the age of 12, and from Land's End to John O' Groats at 15 – he had never cried. On his cycle round the world, he wept on at least four occasions.

At the moment of his greatest personal victory, he gave credit once more to the plan, claiming all he had done in reality was to follow it. Beaumont was proud of the fact he still had some time in the bank, while his profile on Strava, the app that allows like-minded cyclists to track the times and distances of peers, made for typically modest reading. Under Beaumont's circumnavigation, it said simply: 78 rides.

# THE WORLD IS NOT ENOUGH...

## ... you have to abide by a strict set of rules if you want to set an official cycling circumnavigation record (don't forget your antipodal points!)

Fancy your chances breaking the Guinness World Record for circumnavigating the world by bike? Then you'll need to adhere to a strict set of rules. For one thing, the journey must be continuous and the minimum distance travelled no less than 28,968km (18,000 miles). The ride can travel east to west or west to east but, as with Mark Beaumont's choice of the Eiffel Tower, you must start and finish in the same place.

You are also required to pass through at least two 'antipodal' points, those located on precisely opposite sides of the planet. The rules state that while you don't have to start and finish at one of these antipodal points, you must build them into your route. This demands some careful deliberation. For example, a good many of the antipodal points when riding in Australia are located in the middle of the north Atlantic; and those corresponding to North America lie in the Indian Ocean.

The most recent change to the rules came in 2012, when Guinness amended the regulations around flight times. Under the old rules, the clock stopped once a rider reached their departure airport and did not resume until they had cleared customs at their destination.

The 2012 change means the clock doesn't stop at all, so being able to get in and out of airports quickly is vital.

In addition to the rules set down by Guinness, unsupported rides bring additional criteria. These include carrying your own equipment, which seems fair, but also seemingly minor constraints such as only receiving mail at public addresses, a post office for instance. There is also the small matter of isolation; you are not permitted at any point to be joined by friends; in fact, if you do encounter another rider en route, you are advised to maintain a minimum distance of five bike lengths.

Analogies are sometimes made between cycling and the sport of climbing; both often take place in the mountains, require focus and determination to reach the summit, and both are at the mercy of extremes of weather. Climbing can be split, roughly, into two philosophies. The first, 'siege-style' climbing, is characterised by large support teams, who lay provisions along a given route, thus allowing the climbers to focus exclusively on the climb. The advantages, in terms of range and contingency, are huge. The second philosophy, of which there are numerous variations, is best described as 'fast and light'; minimal gear, minimal fuss, the climbers against the elements. While it has an undeniable romance, even glamour, it does leave a lot more to chance. Similarly contrasting views exist when it comes to circumnavigations.

Beaumont's fellow Scot, Jenny Graham, holds the record for the fastest woman to circumnavigate the globe by bike. She completed her 2018 ride in 124 days, entirely unsupported, even resorting to sleeping rough on occasion.

Beaumont himself compares his style of riding to sailing, a sport in which records are lowered frequently by small increments and where detailed planning is key. This approach has a strong precedent in road racing and is best seen in the 'marginal gains' philosophy adopted by Sir Dave Brailsford's Team Sky. In guiding the likes of Bradley Wiggins, Chris Froome and Geraint Thomas to seven Tour de France wins in eight years, Brailsford's idea, put simply, is not to do one thing 100% better but to improve on one hundred things by 1%.

# THE 'BOLD YOUTH' WHO MADE HIS PENNY-FARTHING COUNT

**When Thomas Stevens embarked in 1884 on what would become the first bicycle circumnavigation, he did so with a 127cm front wheel and a revolver.**

The history of circumnavigation by bike has a notably British flavour, but the cyclist who made the first circuit of the world on two wheels could be claimed as an honorary American. When Thomas Stevens set off to travel the world by bicycle, in 1884, the 20th century was some way off but there was the sense of a new dawn, the vast front wheel of the penny-farthing – his bicycle of choice – its rising sun. Cycling was about to enter a boom time. A decade or so later, HG Wells published *The Wheels of Chance* and set loose Mr Hoopdriver, the frustrated draper's assistant who sought freedom from his lowly position aboard his 'safety bicycle'. This design, with two wheels of the same size, was a forerunner of the modern bicycle; the penny-farthing's habit of disembarking its riders at speed from a great height was a principal cause of its demise.

Originally from Hertfordshire, in southern England, the Stevens family had emigrated to the US in 1868. By the early 1870s, they were in San Francisco, on whose vertiginous hills the young Thomas first learned to ride a bike. Exactly what bike it was, history does not record. By 1884, however, he had upgraded, to a jet-black Columbia 'Standard', a penny-farthing whose fearsome 127cm (50in) front wheel was nickel-plated. The manufacturer, the Chicago-based Pope company, was one of America's first, having been in the bicycle business for all of six years.

Packing little more than a raincoat and a revolver, Stevens set off on 22 April 1884. Riding, or 'wheeling' as he called it, along wagon trails and railways, Stevens travelled through the Sierra Nevada, bound for Wyoming. Cycling was already sufficiently popular that he was joined by fellow riders on numerous occasions.

On the first leg of his trip, for example, he was escorted into Laramie by members of the League of American Wheelmen. His ride was widely reported by the mainstream press – the *New York Times* and *Times* of London took a keen interest – but also by a growing number of specialist titles, such long-vanished periodicals as *Adventure Cyclist* and *Outing*, for whom Stevens became a correspondent.

Just over 100 days after leaving San Francisco, on 4 August 1884, Stevens reached Boston. In clocking up a distance of 5955km (3700 miles), he had made the first transcontinental ride across the US; even if, as *Harper's* sniffily pointed out, poor roads and bad weather had required him to walk as much as 1930km (1200 miles).

Stevens spent the winter in New York, bolstering his journalistic career – he would later be dispatched to Africa to join the hunt for the explorer Henry Morton Stanley – finally setting sail for Liverpool on 31 March 1885. Following a 10-day voyage, he departed on the European leg of his ride a few weeks later, escorted from the city by an honour-guard of cyclists. Stevens certainly had a flair for showmanship (his résumé included a role as the manager of London's prestigious Garrick Theatre) and rode through Europe sporting a white pith helmet. On reaching Istanbul, then Constantinople, he improved both his bicycle, with repairs, and his firepower, in the form of a Smith and Wesson revolver.

Having been advised in London to avoid crossing central China, he plotted a course through Persia and across Afghanistan, where he was eventually expelled by the authorities. Stevens found India more agreeable, in part because of the hard top of the Grand Trunk Road, the ancient Asian thoroughfare that runs for 2736km (1700 miles), linking Kabul, Afghanistan, to Chittagong in Bangladesh. He then took a ship from Calcutta to southern China.

Right: promotional poster for Pope Manufacturing, producer of the Columbia 'Standard' penny-farthing that Thomas Stevens rode around the globe. Bottom: illustrations from Stevens' book *Around the World By Bicycle*.

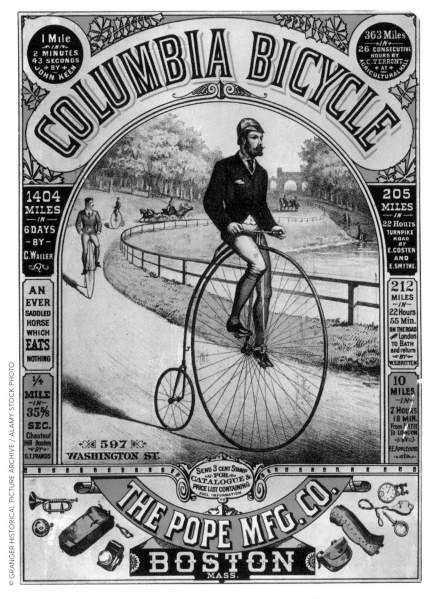

Stevens reached Japan aboard a Chinese steamer, concluding his ride in Yokohama, on 17 December 1886. The book Stevens wrote of his experiences, *Around the World on a Bicycle* (see overleaf), was published in 1887. In the preface, a cycling enthusiast called Thomas Higginson, who had heard Stevens talk at the Massachusetts Bicycle Club, provided a fitting – one might even say prescient – testimonial. He wrote: 'modern mechanical invention, instead of disenchanting the universe, had really afforded the means of exploring its marvels the more surely. Instead of going around the world with a rifle, for the purpose of killing something, or with a bundle of tracts in order to convert somebody, this bold youth simply went round the globe to see the people who were on it.'

Sadly, the 127cm (50in) Standard penny-farthing that had given Stevens such good service met a sorry end. Having carried its master 21,726km (13,500 miles), and through countless exotic locations, it was kept in storage for the best part of 50 years. When the US joined WWII, in December 1941 following the bombing of Pearl Harbor, the bicycle was melted down for scrap as part of the war effort. In recognition of Stevens' ride, the Adventure Cycling Association of America has a Thomas Stevens Society, dedicated to honouring his achievements.

# 'WITH BRAKE IN HAND, I GLIDE SMOOTHLY DOWN THE SLOPE'

**We join Thomas Stevens and his penny-farthing in Persia on the Meshed Pilgrim Road in this extract from Volume II of his Around the World by Bicycle (1887).**

It rains quite heavily during the night, but clears off again in the early morning, and at eight o'clock I take my departure. The whole male population of the village is assembled again at the spot where their experience of yesterday has taught them I should probably mount; and the house-tops overlooking the same spot, and commanding a view of the road across the plain to the eastward, are crowded with women and children.

Four miles of most excellent camel-path lead across a gravelly plain, affording a smooth, firm, wheeling surface; but beyond the plain the road leads over the pass of the Sardara Kooh, one of the many spurs of the Elburz range that reach out toward the south. This spur consists of saline hills that present a very remarkable appearance in places; the rocks are curiously honey-combed by the action of the salt, and the yellowish earthy portion of the hills are fantastically streaked and seamed with white. A trundle of a couple of miles brings me to the summit, from which point I am able to mount, and, with brake firmly in hand, glide smoothly down the eastern slope. After descending about a mile, I am met by a party of travellers who give me friendly warning of deep water a little farther down the mountain. After leaving them, my road follows down the winding bed of a stream that is probably dry the greater part of the year; but during the spring thaws, and immediately after a rain-storm, a stream of brackish, muddy water a few inches deep trickles down the mountain and forms a most disagreeable area of sticky salt mud at the bottom. The streak this morning can more truthfully be described as yellow liquid mud than as water, and both myself and wheel present anything but a prepossessing appearance in ten minutes after starting down its grimy channel.

I am suddenly confronted by a pond of liquid mud that bars my farther progress. A recent slide of land and rock has blocked up the narrow channel of the stream, and backed up the thick yellow liquid into a pool of uncertain depth. There is no way to get around it; perpendicular walls of rock and slippery yellow clay rise sheer from the water on either side. There is evidently nothing for it but to disrobe without more ado and try the depth. Besides being thick with mud, the water is found to be of that icy, cutting temperature peculiar to cold brine, and after wading about in it for fifteen minutes, first finding a fordable place, and then carrying clothes and wheel across, I emerge on to the bank formed by the land-slip looking as woebegone a specimen of humanity as can well be imagined. With the assistance of knife, pocket-handkerchief, and sundry theological remarks which need not be reproduced here, I finally succeed in getting off at least the greater portion of the mud, and putting on my clothes.

A little good wheeling is encountered toward the bottom of the pass, and then comes an area of wet salt-flats, interspersed with saline rivulets. Soon after noon I reach the village of Kishlag, where a halt of an hour or so is made to refresh the inner man with tea, raw eggs, and figs. The weather-clerk inaugurates a regular March zephyr in the east, during the brief halt at Kishlag; and in addition to that doubtful favor blowing against me, the road leading out is lumpy as far as the cultivated area extends, and then it leads across a rough, stony plain that is traversed by a network of small streams. To the left, the abutting front of the Elburz Mountains is streaked and frescoed with salt, that in places vies in whiteness with the lingering-patches of snow higher up; to the right extends the gray, level plain...beyond which lies the great dasht-i-namek (salt desert) that comprises a large portion of the interior of Persia.

# 'I HAVE A PASSION FOR ENDURANCE'

Jenny Graham was in her mid-twenties before she discovered bikepacking – a decade later she was smashing cycling records.

W hen you cycle around the world you pack light: Jenny Graham from Scotland pedalled halfway around the world with just one pair of socks. She departed Berlin on 16 June 2018 with the aim of beating the women's world record for the fastest unsupported circuit, so carrying the minimum of baggage was essential. She arrived back at the Brandenburg Gate 124 days later on 18 October having cycled 28,970km (18,000 miles) across four continents and 16 countries, averaging 251km (156 miles) per day and beating the previous record holder by 20 days.

**Why did you decide to cycle around the world?**
I had my son, Lachlan, when I was 18 so my twenties were about family life. But when I got into my mid-twenties I discovered outdoor activities like hillwalking, skiing and bikepacking. I always really enjoyed being in the hills. Biking was a regular thing I did but I was never obsessed about it. Not at the start.

I had a passion for endurance. When I met some people in Inverness, I started cycling more miles. We'd do 100-mile back-to-back days every weekend. It had me questioning how much further I could go. I was always setting challenges like Land's End to John O'Groats in four days.

Then I went on a training camp with the Adventure Syndicate [a female-cyclist collective], which I'm now a director of, and met a coach called John Hampshire. He offered me a year's free coaching and I knew that I wanted to do something amazing with it. I was looking at races but the round-the-world record kept appearing. I think I sat with the idea for a couple of months because it was such a big deal to say out loud and actually believe you might be capable of cycling around the world.

© JAMES ROBERTSON

### Why self-supported?

It's just the way I like to travel. It's so simple. You're only carrying what you need to fix your bike, fix yourself, and to be able to sleep at night and eat. And that's it. No extras. I remember in Australia looking at buying another pair of socks. I went on one pair of socks halfway around the world. But my feet started getting cut up because of the dust in Mongolia. I thought about it for two days because it was such a big deal!

The challenge isn't just about putting the miles in. It's about keeping strong enough to make good decisions when you find yourself in situations that you need to navigate through. The interactions you have are very different if you're self-supported and that was a huge part of the trip for me.

I didn't have any luxuries. The one piece of kit that I couldn't have done it without was my dynamo hub. It meant that I always had lights and it could charge my GPS and battery packs.

### How did you plot your route?

I spoke to Mark Beaumont before I left. He had just set the male supported record and had had a big team planning a route. Some continents are straightforward. There's one fast way to cross Australia. It's the same with crossing Canada and America to get the most miles. In Europe, you can go across Russia and down into Mongolia, which Mark had done, or you can head a bit further south and either go through the 'stans and India or do what I had planned and cross Kazakhstan to head into the west of China towards Beijing. Mark had looked very carefully at the winds and I decided to take his

route across Russia. Maybe it was faster but the roads in Russia were so out-of-this-world dangerous that I would never do that again. For about 1200 miles east of Moscow it was so crazy that I started riding at night.

### How did you navigate?

Some countries were easy. In Australia there's just one road and then you turn left at Melbourne. In Canada you're on the same highway for a couple of weeks. When you got into cities or in places like China, it was much harder because you couldn't read any of the signs so you were relying on having your electronics charged up and your route to be right. I'm quite good at winging it. I got lost a few times, mostly because I was tired and hadn't looked at my GPS at a roundabout. If people on Twitter saw that I was off-route on my tracker – and I'd sometimes have to go off-route to get food or sleep – they would send me directions.

### Was it memorable meeting locals?

The section across Mongolia and into Siberia and western China blew me away: the hospitality and their surprise at seeing me. They would bend over backwards, even though I was there for a fleeting moment. It really hit me how privileged I was to be able to do this.

I definitely noticed the difference with Western culture. I found that in Asia people wanted your basic needs to be met and it didn't matter how little they had. One night in Australia I'd just arrived in Adelaide. It was 2am and raining. I'd gone to the door of this hotel like a drowned rat and asked if I could have a room

and the response was 'no, we've closed the tills'. And he shut the door in my face. As Westerners we can let go of the important stuff.

### What were your favourite places?
There was never a place where I wished that I could stay. I was always relieved that I didn't have to make that decision. Mongolia and China were beautiful and I loved the culture. And there's so much space in Alaska and Canada – and wild animals. Bear, moose and bison wandered onto the road. I saw bears and I was petrified. I was sleeping out in a bivvy bag. But now that I'm home I can say that they were gorgeous.

### Were you surprised by what your body can do?
Yes, absolutely. I thought that I've got to seize up at some point. But it was more surprising what your mind can do. Your legs keep going. Once you can ride for a few hundred miles back to back and you know that you can do

that, then your legs can do anything, But your mind, that will stop you.

### What did you eat?
I had beautiful fresh food going through Mongolia. The worst was having to live out of petrol stations and fast food places. Coffee was my thing. I would plan where I was going to sleep to be as close to the coffee as I could be the next morning. I knew that it would get me out of bed.

### What did you learn about yourself?
This voice in the back of my head was saying 'you're going to mess this up!' by doing something like lose my bank cards. But once I had survived Moscow and finished that leg at the Great Wall of China, there was a sense that I could do this. That was a lovely feeling that I was glad I had at points later in the trip. Things might seem like massive barriers but you can problem-solve your way through anything.

Clockwise from top left: heading east through Poland; starting from the Brandenburg Gate in Berlin; preparing for the day after bivvying overnight; pedalling into the night.

61

# OFF THE ROAD AGAIN

Cycling north from Patagonia, Ben Page realised he preferred riding when the asphalt ran out – what followed was an epic, off-road odyssey across the planet's great wildernesses.

© BEN PAGE

Most cyclists planning to ride around the world look for good roads. Many accept there will be times when they must endure bad ones. Very few would plan a route that involved no roads at all. One, in fact. That cyclist is Ben Page. By his own admission, the young Englishman is terrible at planning. All Page knew was that, when he started in 2014, he wanted to ride overseas; on four of seven continents, for a minimum of three years.

Born in 1992, Page had, as a child, been a promising road racer, good enough to be considered for a professional contract. Knowing he was a strong, if disorganised rider, he decided to test his fitness on expedition-style sorties. Aged just 18, he rode from his home, in Bristol, to the French Alps in just seven days, clocking 240km (150 miles) a day – and felt fresh as a daisy afterwards. He cycled across America in two-and-a-half months.

When he set out in December 2014, aged 22, what semblance of a plan there was involved heading for Europe. Things took a swerve when a group of friends, off to ride through Patagonia, invited Page to join them. With the Patagonian trip done, Page pondered his next move. Reasoning he was at the foot of South America, the most accessible riding available – living off the remnants of a student loan, flights were strictly limited and hotels were out altogether – lay to his immediate north. So, Page rode across the entirety of South, Central and North America.

It was an approach he adopted for each subsequent continent he covered; bottom to top or left to right. Page simply fired up the GPS on his smartphone and followed the arrow, road or no road. In fact, as he rode he realised he actively preferred riding off-road, so decided to develop his routes accordingly. That required a change of vehicle. Twelve months into the trip, while cycling across Canada, Page's various pleas to potential sponsors bore fruit, in the shape of Fatback Bikes, an Anchorage-based manufacturer of extreme off-road mountain bikes. Agreeing to support him, Page ditched his trusty tourer and beefed up his rig to a carbon-fibre beast, complete with full bikepacking kit. He also bought a camera.

The film he made of his time in the Canadian Arctic, *The Frozen Road*, became an award-winning documentary, its opening credits (eventually) adorned with the palm leaves of critical acclaim. While Page's Instagram feed features sumptuous shots of the Northern Lights, there are scenes in *The Frozen Road* that depict the less desirable aspects particular to his trip. Namely, the relentless stream of decisions required when you're making it up as you go along.

From Canada, Page flew to Beijing, before off-roading all the way to Istanbul. En route, he slept in yurts, forded rivers with the help of Mongolian goat-herders and slept beneath the stars in Uzbekistan's remote Karakalpakstan Desert. Yet another decision, however, awaited him in Turkey. Ride home for Christmas – by now it was November 2016 – or head for

© BEN PAGE

Clockwise from below: Ben Page riding a frozen road through northern Canada; crossing a river in Mongolia with the help of a horseman; enjoying the sunset in Africa.

© BEN PAGE

Cape Town and pedal his way across Africa? He chose the latter, of course, taking a not-so-classic Cape to Cairo route, off-road, and arriving back home three years after his departure.

It is an oft-aired complaint that technology makes the planet smaller, perhaps even too small. Page might say precisely the opposite: that thanks to readily available GPS – his preferred app was maps.me – and some high-tech bike engineering, there is as much of the world to explore as ever.

## VIEW FROM THE CHEAP SEATS

When riding around the world off-road, the intrepid Ben Page set himself a budget of around $4.50 (£3.50) a day, excluding extras such as flights. Or the camera he purchased in North America, and which helped him become a multi-award-winning documentary-maker. Purist backpackers, or bikepackers in this case, pride themselves on spending as little as possible on daily outgoings. In 2016, the British adventurer Laura Bingham claimed to have ridden 6920km (4300 miles) across Ecuador and Argentina without spending a single cent. She traded spare clothes for accommodation, worked on farms for food, or scouted for leftovers. Failing that, she relied on the kindness of strangers.

# ON YOUR MARKS, GET SET...GO FOR IT!

If you don't have the time/money/endurance to cycle around the entire globe, you could try breaking it down, one continent at a time. Four of the planet's seven continents host coast-to-coast ultra-endurance races, mass-participation events known as 'sportives'; and although not technically a race, Africa has at least one continent-wide event, the Tour d'Afrique. Asia is a bit more complicated but perfectly doable with a little planning. What are you waiting for?

### THE DADDY: THE RACE ACROSS AMERICA

The 4800km (3000-mile) Race Across America (RAAM) is rightly acknowledged as the most gruelling bike race in the world. Competitors with an eye on victory will ride for 23 hours a day, for around eight days solid. The fastest winning time was set in 2014, by Austria's Christoph Strasser, who clocked an average speed of 26.43km/h (16.42mph), completing the course in seven days, 15 hours and 56 minutes. The cut-off for slower riders is 12 days; unlike the Tour de France, there is no 'broom wagon' to sweep up lagging cyclists, so the 50% of starters who don't finish will climb back into their own support vehicles. Sleep deprivation means hallucinations are common. In 2017, for example, two-time finisher Jason Lane believed his support team were trying to kidnap him and thus refused their offers of food and water.

First run in 1982 (as the Great American Bike Race), the RAAM travels west to east non-stop, from California to Maryland. And while that's 965km (600 miles) longer than the average Tour de France, which is run over three weeks, RAAM riders aim to complete the course in a third of that time. The fastest allow themselves just 30 minutes off the bike in a given 24 hours, while the most ambitious will ride from California to Kansas before stopping for their first break, a distance of 1450km (900 miles). Support cars are packed with spares, water and food – and saline drips. Riders start at intervals and techniques common in other races, such as drafting – tucking in behind another bike for an aerodynamic advantage – are prohibited. And whereas a Grand Tour rider will launch an 'attack', a bid to gain time, on a single climb or road, RAAM contenders might increase their speed incrementally, across an entire mountain range, even an entire state. In addition to seeing things and hearing voices, other downsides include saddle sores, which develop after a couple of days. Some consolation lies in the landscape: the Arizona desert, Colorado and the Midwest.

### THE CLASSIC: THE TOUR D'AFRIQUE

Depending on your cycling preference, the Tour d'Afrique (TDA) can be ridden as either an expedition or a race. For those minded to

Left: cycle past Cairo's pyramids on the Tour d'Afrique. Right: the gravel roads of Peru's backcountry on the IncaDivide route.

© AXELLE | BAUER-GRIFFIN | GETTY IMAGES

© PUTT SAKDHNAGOOL | 500PX

© CAMILLE MCMILLAN

explore the climatic diversity of the continent, it offers a spectacular route that showcases Africa's history and landscape.

The TDA is run annually, between January and May, when the climates are optimal in the 10 countries it passes through. Allowing for slight variations – if there is political unrest, for example – the TDA features around 100 stages, and these vary in length between 40

Clockwise from above: the Race Across America reaches San Diego, California; Utah's Monument Valley, one of the RAAM's highlights; epic scenery on the Transcontinental race.

## THE TRANS-CONTINENTAL

'I left Grindelwald by a singletrack road, closed to all traffic but the yellow post buses, whose three-note horn heralded their appearance around the hairpins as I climbed higher. The grass, not long emerged from the winter snows, sparkled with tiny flowers, and placid brown cows serenaded me with their bells.'
*Emily Chappell*

and 100km (25 and 62 miles). The emphasis is as much on experience as it is on speed. Riders are encouraged to cycle together and support vehicles are provided. That said, to complete the course you will still need to pedal your way through a hefty 12,070km (7500 miles).

The route starts on the banks of the Nile, where riders must negotiate the crowds of Luxor and Aswan, before the road opens out in the deserts of Sudan. Next comes a traverse of the Simien Mountains, in Ethiopia, while highlights in Kenya include remote Dida Galgalu. From Tanzania, riders cross into Zambia, where they visit Victoria Falls. In Botswana, the relief of flat roads is tempered by the presence of roaming elephants. The peloton then skirts the Namib Desert before it enters South Africa, bound for Cape Town.

The record for the fastest time to complete the distance was set in 2003, when nine riders reached Cape Town after 100 days on the road. The time still stands as a Guinness World Record for the swiftest crossing of Africa by bicycle.

## THE DIRTBAGGER: EUROPE'S TRANSCONTINENTAL

It may not be as long as the Race Across America but Europe's Transcontinental has a certain wildness to it, a dirtbag-climbing vibe. Totally unsupported, the rules dictate that riders must carry all their own kit and while most compete with a bivvy bag and a camping mat, few take a tent, so the ability to sleep by the roadside is important.

The race was founded by the late British endurance rider, Mike Hall, with the first edition run in 2013. The route, from northwest to southeast Europe, varies each year, as does the

# RACING ALL OVER THE PLANET

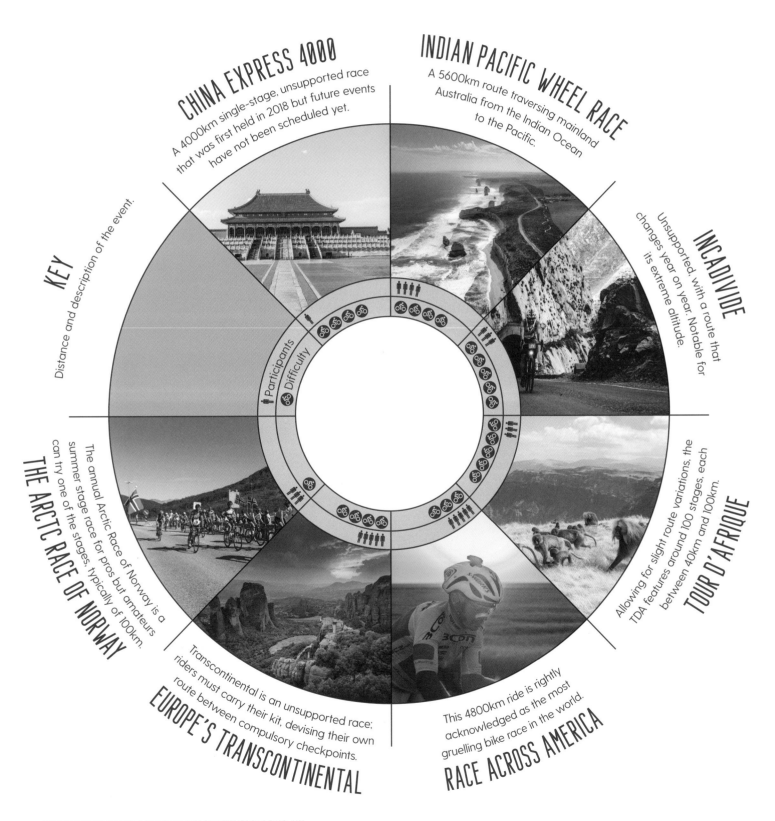

**CHINA EXPRESS 4000**
A 4000km single-stage, unsupported race that was first held in 2018 but future events have not been scheduled yet.

**INDIAN PACIFIC WHEEL RACE**
A 5600km route traversing mainland Australia from the Indian Ocean to the Pacific.

**KEY**
Distance and description of the event.

Participants
Difficulty

**INCADIVIDE**
Unsupported, with a route that changes year on year. Notable for its extreme altitude.

**THE ARCTC RACE OF NORWAY**
The annual Arctic Race of Norway is a summer stage race for pros but amateurs can try one of the stages, typically of 100km.

**TOUR D'AFRIQUE**
Allowing for slight route variations, the TDA features around 100 stages, each between 40km and 100km.

**EUROPE'S TRANSCONTINENTAL**
Transcontinental is an unsupported race; riders must carry their kit, devising their own route between compulsory checkpoints.

**RACE ACROSS AMERICA**
This 4800km ride is rightly acknowledged as the most gruelling bike race in the world.

Clockwise from top, scenes from the IncaDivide race: valley roads; the start; local Peruvians; riding into the Andes.

distance, typically between 3200 and 4000km (2000 and 2500 miles). Originally run between London and Istanbul, for several years the race start was the Belgian city of Geraardsbergen, its finish Meteora, in Greece. The 2019 edition was the first to run east to west, from Burgas on Bulgaria's Black Sea coast, to Brest in northern France. Variations in distance have produced a range of winning times, typically between seven and 10 days.

Interest in the race has grown quickly and it's probably fair to say that, among millennials, it already enjoys a certain cult status. Just 30 people rolled out for the inaugural race, yet by 2016 that number had grown to 600 applications for 300 places; over 250 started the 2019 event.

Riders are issued with 'brevet' cards, a detail borrowed from the French discipline of brevet riding, whereby competitors are required to have their cards stamped at set checkpoints. Otherwise, riders are free to choose their own routes. Drafting of any kind, behind cars as well as other bikes, is forbidden.

Riders are also issued with GPS trackers, which means the Transcontinental is notable for its addition to the relatively new cycling phenomenon of 'dot-watching', following the progress of individual riders online from the comfort of an armchair.

Below: a Peruvian pitstop on the IncaDivide.

## THE NEWBIE: THE INCADIVIDE

The southerly cousin of the Race Across America is one of the babies of the endurance sportive family. First staged as recently as 2017, the IncaDivide owes its name, in part, to the Tour Divide, the annual mountain-biking endurance race that traverses the Canadian Rockies and ends at the border with Mexico. The brainchild of French endurance rider Axel Carion, the race is overseen by Carion's company, BikingMan. Something of a cycling aficionado in the region, Carion holds the record, along with Andreas Fabricius, for the fastest crossing of South America; the 49 days, 23 hours and 43 minutes it took the pair to ride from Cartagena in Colombia, to Ushuaia in Argentina.

Like the Transcontinental in Europe, the IncaDivide is unsupported and the start and finish locations also alternate; the first race, for example, started in Quito, in Ecuador, and ended in the Peruvian city of Cuzco, a distance of 3540km (2200 miles), which the winning rider completed in 17 days. What marks out the IncaDivide from its rivals is its extreme altitude; much of the original route followed ancient Inca trails, crossing the Andes via numerous passes in excess of 4000m (13,000ft). In fact, 2019 saw the race rise to new heights, thanks to a staggering, cumulative elevation of 32,500m (106,600ft).

© YUNSUN_KIM / SHUTTERSTOCK

(The 2019 Tour de France, by comparison, the highest in its history, featured just six climbs over 2000m/6560ft.) A (very) highlight of the route was the Huascarán National Park, at 6768m (22,200ft). For good reason, that year's IncaDivide was restricted to 1600km (1000 miles) in one country, Peru. It featured just three mandatory checkpoints and a 10-day cut-off.

### THERE AND OUTBACK AGAIN

Another relative newcomer – like the IncaDivide it was first held in 2017 – the Indian-Pacific Wheel Race has sound historical provenance. The 5600km (3500-mile) route traverses mainland Australia, from west to east, the Indian Ocean to the Pacific, from Fremantle to the Sydney Opera House. In so doing, it crosses the arid Nullarbor Plain, passes through the winelands of South Australia, and takes in the coastal sweep of Victoria's Great Ocean Road, before rising over the Australian Alps, in the southeast of the continent.

The race is inspired by Australia's first endurance riders, known as 'overlanders', who at the turn of the 20th century set their machines rolling through the outback in a bid to become the first, and later the fastest, to cycle the great distances that separated

### MIKE HALL

In 2017, 500km from the end of the inaugural 5500km (3418-mile) Indian-Pacific Wheel Race in Australia, Mike Hall was killed by a driver before dawn near Canberra. After completing his first self-supported race, the Tour Divide, in 2011, he returned to win that race twice and the Trans Am Bike Race. He founded the Transcontinental Race in 2013.

Australia's cities. The overlanders became minor celebrities, warmly greeted as they passed through out-of-the-way rural towns. To maintain that historic relationship between town and country, the race passes slap-bang through the middle of such major urban centres as Adelaide, Melbourne and Sydney. So, a rider who the previous night might have bedded down with no more company than the Southern Cross, can the next day find themselves in the middle of the morning commute. It's a nice touch. Indeed, the Indian-Pacific Wheel Race has done well to survive some testing times. In its inaugural year, the race was suspended after Mike Hall, a leading light in the ultra-endurance cycling community, was killed by a car (see left).

### OUT IN THE COLD: ANTARCTICA

It won't come as a huge surprise to discover Antarctica doesn't host a transcontinental cycling event. But that doesn't mean cyclists have abandoned hope of staging one, or at least riding significant parts of it. In 2013, the British adventurer Maria Leijerstam – aka the 'Ice Machine' – established a record for riding from the coast of Antarctica to the South Pole. Leijerstam's mark of 800km (500 miles), in a time of 10 days, 14 hours and 56 minutes, may

© DAVID MERRON PHOTOGRAPHY / GETTY IMAGES

seem slight when compared to the mileage clocked up during some of the other events featured on these pages. Leijerstam, who rode a specially adapted recumbent bike, can point to some mitigating factors. The first part of the ride includes the ascent of a 1000m (3280ft) mountain range, where temperatures of -40°C (-40°F, provided you go in summer) are exacerbated by strong winds.

One fellow cyclist whose competitive instinct was piqued by Leijerstam's achievement is none other than British Olympic track legend, Sir Chris Hoy. In 2017, Hoy announced plans to attempt the coast-to-pole record the following year but was forced to delay. Used to muscling his way around the track in minutes, if Hoy ever does reach Antarctica, he will find a very different kind of sporting arena. On Leijerstam's ride, the veteran adventurer consumed up to 4000 calories a day, the minimum required to power her for spells of up to 17 hours in the saddle – Leijerstam still managed to lose around 8% of her body weight.

## JOIN THE DOTS: DIY ASIA

The sheer size, as well as the geographical and political complexities of Asia, mean it does not have a single, transcontinental event. The China Express 4000, a single-stage, unsupported race first held in 2018 will, in theory, get you from Ürümqi to Beijing, in a little under 4000km

(2500 miles). Information on future events is hard to come by, however, perhaps due to the ongoing political tensions in its starting city, which is also the capital of the Xinjiang Uighur Autonomous Region. The 'dot-watching' site www.trackleaders.com, which monitors riders by GPS, suggests the 2018 event may have had as few as two participants. Elsewhere, BikingMan, organiser of the IncaDivide, offers a 1100km (680-mile) event through the mountains of Taiwan. Further east, the Japanese Odyssey covers 2500km (1550 miles) of the archipelago's roads, with checkpoints along the way but apparently very minimal time requirements. So, basically a tour.

Right: Maria Leijerstam reaches the South Pole. Below: Adélie penguins, Antarctica. Opposite: the Indian-Pacific Wheel Race concludes at the Sydney Opera House.

© MARIA LEIJERSTAM

# PLANE
## BY

O n the morning of 6 April 1924, a team of clean-cut US Army pilots climbed aboard three planes at a Seattle airfield. As they coaxed their aircraft into the sky, they were embarking on a journey that would change the nature of travel. In their magnificent, specially commissioned flying machines, the pilots of the 'World Flight' aimed to make the first circumnavigation of the globe by air. Their success, six months later, would pave the way for aviation adventurers and humble holidaymakers alike.

From the moment Orville and Wilbur Wright made the first powered flight in 1903, it became the dream of aviators to push the boundaries of life aloft; to fly first from one continent to another, and then around the world. An initial transcontinental flight, across North America, had taken place in 1923. The first crossing of a major ocean, the Atlantic, occurred in May 1919, albeit in a flying boat. The maiden non-stop transatlantic flight came a month later, in the shape of British aviators John Alcock and Arthur Whitten Brown. On 14 June 1919, flying a modified ex-RAF Vickers bomber, Alcock and Brown travelled from St John's, in Newfoundland, to Ireland, a distance of 3040km (1890 miles).

These were exciting times; as pilots from around the world sought glory in the skies, their exploits drew wealthy, ambitious, adventurous characters into aviation's increasingly glamorous orbit. One was Raymond Orteig, a French-born New York hotelier, whose Orteig Prize offered a $25,000 reward for the first successful, non-stop transatlantic flight between New York City and Paris. The high stakes attracted high-profile contenders, many of whom soon fell to earth. In September 1926, French flying ace René Fonck's Sikorsky S-35 crashed on take-off at New York's Roosevelt Field. In April the following year, in Virginia, US Navy pilots Stanton Wooster and Noel Davis were killed as they prepared to make their bid; a month later, *L'Oiseau Blanc*, a seaplane piloted by French war heroes Charles Nungesser and François Coli, disappeared off the coast of Ireland, shortly after taking off from Paris.

Which explains why the 5790km (3600 miles) flown by Charles Lindbergh in his *Spirit of St Louis*, in 1927, made front pages across the globe. The first solo non-stop transatlantic flight, it was also almost twice the distance flown by Alcock and Brown. The period between the wars is generally regarded as the golden age of aviation. Air races and record-setting flights were big news. Aeroplanes evolved from wood-and-fabric biplanes to streamlined metal monohulls. The military's embracing of air power also helped aviation come of age. And the records kept racking up. In 1935, the Curtiss-Robin biplane *Ole Miss* established a world record for sustained flight; using air-to-air refuelling, it stayed airborne for 27 days, despite severe thunderstorms and an electrical fire in the cabin.

Leading the charge was Howard Hughes. The business magnate already had a clutch of records to his name when, on 14 July 1938, he completed a flight around the world in just 91 hours, beating the previous record, set by Wiley Post in 1933, by almost four days. Taking off from New York City, Hughes continued to Paris, Moscow, Omsk, Yakutsk, Fairbanks and Minneapolis, before finally returning to New York. Accompanied by a four-man crew, Hughes flew a Lockheed 14 Super Electra NX18973, a twin-engine transport fitted with the latest radio and navigational equipment. It's evident that he wanted the flight to be a triumph of American aviation technology, illustrating that safe, long-distance air travel was possible. By 1939, Hughes was majority shareholder in Trans World Airlines (TWA).

In commercial aviation at least, Hughes was a relative latecomer. TWA was founded in 1930 but the history of the industry goes back significantly further. The world's first airline was established in Frankfurt in 1909, with a fleet made up entirely of Zeppelins. Aircraft Transport and Travel, the earliest fixed-wing airline, was founded in Britain by George Holt Thomas in 1916 and would eventually go on to become British Airways. A few years later, the first two of the world's four oldest surviving airlines were set up: KLM and Avianca were founded in 1919; Qantas followed in 1921; and Czech Airlines got underway in 1923, a year before the World Flight's pioneering expedition.

The first scheduled around-the-world service for paying passengers was launched by Pan American World Airways, the now defunct Pan Am, in 1947. A passenger boarding its elegantly named Flight 001, in San Francisco, could look forward to stops in Honolulu, Hong Kong, Bangkok, Delhi, Beirut, Istanbul, Frankfurt, London and finally New York. An eastbound service, Flight 002, originated in New York and stopped in the same cities. An economy-class ticket cost around $2000 but was economy in name only; sumptuous furnishings, fine wine and gourmet food were all served as standard.

Above: John Alcock and Arthur Whitten Brown take off from Newfoundland in June 1919 on their maiden transatlantic flight. Left: the 'Ace of Aces', René Fonck.

Left: Charles Lindbergh in 1927 with his *Spirit of St Louis*. Below, from left: aviation record-breaker Howard Hughes in 1947; Charles Nungesser and François Coli's ill-fated transatlantic attempt commemorated in a 1935 Liebig collectors' card; stamps celebrating the achievements of Lindbergh, and Alcock and Brown.

The glamour continued into the 1950s. A flight, particularly overnight, was envisioned as a cocktail party in the sky, and passengers attired themselves accordingly in their finest evening wear. Champagne and brandy flowed freely; for today's packet of peanuts or a limp sandwich, read lobster; instead of various concoctions hastily reheated (chicken or beef?), think roast beef carved at your table. The Boeing 747s and Airbus A300s that are currently in service may be bigger and faster, yet compared with Pan Am's opulently upholstered Lockheed Constellations, they feel distinctly utilitarian.

# HOW AMERICA SOARED

After WWI, Americans saw powered flight as the preserve of daredevils and the military. So what could be done to show that aeroplanes might connect the world? Fly around it.

From the moment the three specially commissioned Douglas World Cruisers left Seattle, in April 1924, the World Flight pilots would not see home again for 175 days, by which time they had covered over 42,398km (26,345 miles) and made an astonishing 74 stops. The planes were named for American cities: *Chicago*, *Boston* and *New Orleans*. A fourth plane, *Seattle*, had been experiencing mechanical problems and so there was little option but to delay its take-off, in the hope that it would catch up with the World Flight squadron en route.

## A NOBLE, GLOBAL VISION

For all the daredevils competing to notch aeronautical firsts, as a form of mass transportation in the US, the aeroplane was, by 1924, barely more popular than its more basic predecessor, the airship. True, the first airlines had been founded (the very first being all Zeppelins) but they were in Europe; it would be three years before Lindbergh made it non-stop across the

Atlantic and, stateside, the general public had yet to be convinced. The Douglas World Cruisers fitted out for the World Flight had originally been torpedo bombers, confirming suspicions that the true calling of aeroplanes was military.

There were many in the fledgling Army Air Service who concurred, yet only by a high-profile show of aerial prowess would the government persuade the public of the plane's wider applications; the broader objectives were, notionally, humanitarian and political. Just as, in 1907, Theodore Roosevelt's Great White Fleet (see p.36) had circled the globe to bolster support for a US Navy, so the World Flight would create popular demand for the Army Air Service and showcase America's rapidly expanding aeronautical industry in the process. It was no coincidence that the first US airline, the Western Air Express service connecting Salt Lake City and Los Angeles, took to the skies two years later, in 1926. That same year, the Army Air Service became the US Air Force.

## PLOTTING THE COURSE

Planning the World Cruisers' route was a massive

Top: President Calvin Coolidge inspects one of the four Douglas World Cruisers, the first planes to fly around the world.

Opposite, clockwise from top left: Charles Kingsford Smith, who piloted the maiden transpacific flight in 1928; Kingsford Smith and the *Southern Cross* touch down after the epic flight from the US to Australia; the Douglas World Cruiser *Boston*.

undertaking. The critical task of establishing remote supply-and-repair depots, on both land and sea, called upon a host of federal services. The US Army Air Service, Navy, Coast Guard and Bureau of Fisheries all shared the responsibility of distributing thousands of gallons of fuel and numerous spare parts – including 35 spare engines – across the globe, in many cases to locations where planes had yet to fly.

To keep their craft light, each pilot was limited to 135kg (300lb) of supplies. That meant tough decisions. Alarmingly, they took neither parachutes nor life jackets, although it later transpired that Lieutenant Leslie Arnold, pilot of the *Chicago*, had hidden a life jacket in his plane's bulkhead. This careful planning would prove crucial, greatly increasing their chances of success. During the six-month mission, at a series of prearranged waypoints, each World Flight aircraft had its engine changed a total of five times and new wings fitted twice.

## CLEARED FOR TAKE-OFF
From their starting point in Seattle, the pilots flew up the coast of Canada to Alaska, where they faced freezing temperatures, along with sudden, violent storms. By far the greatest danger was the thick, unpredictable fog – it was this that did for the beleaguered *Seattle*; pushing its engine to the limit as it sought to catch the main group, on 30 April it ran into heavy fog near Port Moller, crashed into a mountainside, and was destroyed. Fortunately, the pilots survived.

The World Flight's crossing of the Bering Sea, from the Aleutian to the Komandorski Islands, on 15 May was the first flight across that part

of the Pacific. (The first transpacific flight, from the US to Australia, was only made in 1928, by Charles Kingsford Smith). It was in the Aleutian Islands that the pilots encountered williwaws, which they dubbed 'woolies', strong winds that tore down from the mountains at speeds of up to 120km/h (75mph).

Though authorised to land on the Soviet-held Komandorskis, the World Flight had to skirt the continental Soviet Union, which had not permitted them to fly through its airspace.

## JAPAN TO SOUTHEAST ASIA
In Japan they found welcoming, albeit circumspect, hosts. Excited by the prospect of pioneering aviation, but not necessarily by the American military, the Japanese government insisted the flight take a serpentine onward route in order to protect Japan's military secrets. The rivers and harbours of China, on (➔ p.82)

(➔ p.82)

## ENGINE OF CHANGE
Designed in 1917 for the US government, which wanted a standard engine in 4-, 6-, 8-, and 12-cylinder versions to mass-produce for combat aircraft, the Liberty was America's most important contribution to WWI aeronautical tech. The 20,748 Liberty 12 units that were produced were in widespread use until the 1920s and 1930s.

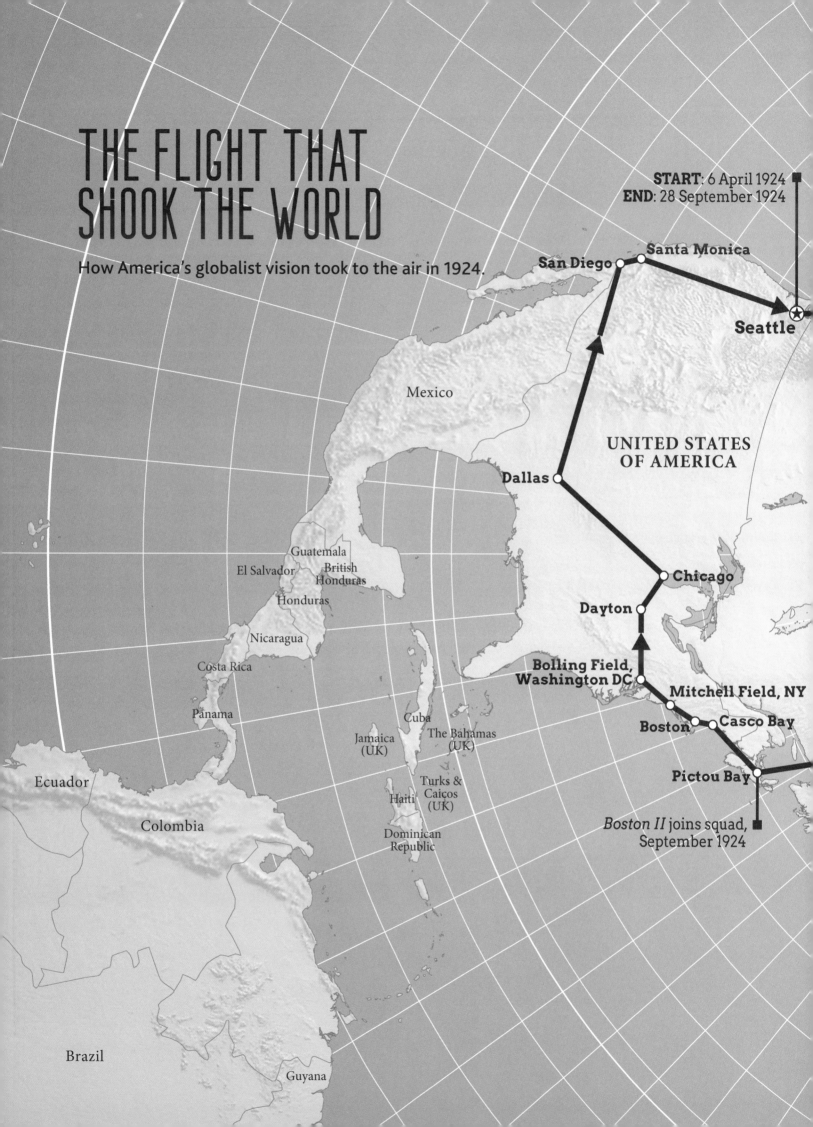

# THE FLIGHT THAT SHOOK THE WORLD

How America's globalist vision took to the air in 1924.

**START**: 6 April 1924
**END**: 28 September 1924

Santa Monica

San Diego

Seattle

Mexico

UNITED STATES
OF AMERICA

Dallas

Guatemala
British
Honduras
El Salvador
Honduras
Nicaragua
Costa Rica
Panama

Chicago

Dayton

Bolling Field,
Washington DC

Mitchell Field, NY

Boston · Casco Bay

Cuba
Jamaica (UK)
The Bahamas (UK)
Haiti
Turks & Caicos (UK)
Dominican Republic

Pictou Bay

*Boston II* joins squad,
September 1924

Ecuador

Colombia

Brazil

Guyana

*Seattle* crashes,
30 April 1924

*Aleutian Islands*

**Atka**

**Attu Island**

**Kasumigaura**

**Osaka**

**K**

**Dutch Harbor**

**Chignik**

**Port Moller**

**Yetorofu**

Japan

**Aomori**

**Kanatak**

**Nikolskoye**

**Paramushiru**

Sakhalin

Korea
(Japan)

**Shan**

rince
upert

**Seward**

**Sitka**

Alaska
(USA)

Manchuria

North Pole

Mongolia

CANADA
(UK)

UNION OF
SOVIET
SOCIALIST
REPUBLICS

Greenland
(Denmark)

y Tickle   **Ivigtut**   **Ammassalik**

**Fredricksdal**

Iceland

**Reykjavik**

**Hornafjord**

Finland

Sweden

**Thorshavn**

Forced landing and
capsizing of *Boston*
3 August 1924

Norway

Estonia

Latvia

**Kirkwall**

Lithuania
East
Prussia

White
Russia

United
Kingdom

Denmark

**Brough**

Germany

Poland

Ukraine

**Band**

Ireland
(UK)

Netherlands

Azerbaijan

Georgia

**London**

Belgium

Czechoslovakia

Rumania

Armenia

**Vienna**

**Budapest**

Arrival in Paris:
Bastille Day,
14 July 1924

**Paris**

**Strasbourg**

Austria

Hungary

**Bucharest**

**Bagdad**

Switzerland

Turkey

Syria

Yugoslavia

Bulgaria

Kuwa

France

**Constantinople**

**Konia**

**Aleppo**

Iraq
(UK)

Portugal

Spain

Italy

Albania

Greece

Cyprus
(UK)

Lebanon
(France)

Trans-Jordan
(UK)

Sau

Far left: New York celebrates the completion of Charles Lindbergh's 1927 transatlantic flight with a ticker-tape parade. Top: the Wright brothers' 1909 Military Flier during trials at Fort Myer, Virginia. Left: aeronautical icons Orville and Wilbur Wright.

Above: Lieutenant Lowell Smith checks the engine of Douglas World Cruiser *Chicago*, a task that cost him a broken rib in Calcutta during the 1924 circumnavigation. Right: 1927 photomontage of Lindbergh's *Spirit of St Louis* over Paris.

the other hand, proved particularly chaotic, a stressful experience for both pilots and ground crew. With the lead plane, *Chicago*, routinely experiencing mechanical problems, it was considered prudent to send it ahead. Moored in Shanghai's busy harbour, *Boston* and *New Orleans* found themselves hemmed in by so many small fishing vessels, they took off a full day after *Chicago*.

Malfunctions continued to dog the *Chicago* and she was forced to land repeatedly. On one occasion, in the Gulf of Tonkin, locating her proved sufficiently difficult that the crews of *Boston* and *New Orleans* put down themselves and took up the search by boat. They found *Chicago*, in the early hours, her crew enjoying a drink with some local sailors who had happened upon the stricken aircraft. It took 10 hours to tow the *Chicago* to the port of Hué in French Indochina (now Vietnam), some 25 miles away. Other obstacles centred not so much on engineering but etiquette. In Saigon, Indochina's capital, the pilots found they could not get served in the city's finer restaurants thanks to a lack of dinner jackets; even the loan of shirts and trousers from their naval support vessels were deemed sartorially inadequate.

## INDIA TO EUROPE

Misfortune was to strike the *Chicago* once more, in Calcutta; while inspecting the aircraft, Lieutenant Lowell Smith slipped and broke a rib – he nevertheless insisted on completing the mission. The World Flight chuntered on its way, across the tropical jungles of the Indian

## BARNSTORMERS AND REBELS

The pilots of the *Chicago* evoked the wild world of early aviation. **Lieutenant Leslie Arnold** just missed seeing combat in WW1. Instead, he travelled America as a military 'barnstormer', generating public enthusiasm for aviation. **Lieutenant Lowell Smith** served briefly as the engineering officer in Mexican revolutionary Pancho Villa's three-plane 'air force'. After WW1, Smith led the first aerial fire patrol in the Pacific Northwest.

subcontinent, and on through the blasting sands of modern-day Iraq and Jordan. It was here the pilots allowed Linton Wells, an Associated Press photographer, to join them for part of the flight; it was Wells who captured the now-famous shot of Lieutenant Leslie Arnold, seeming to climb from the cockpit mid-air, as the World Flight crossed into Europe. The US aviators reached Paris on Bastille Day, 14 July 1924, and from there progressed to London and then the north of England, in preparation for the first leg of the Atlantic crossing, to Iceland.

## ALL AT SEA

It was over the Atlantic that the World Cruisers would be tested to their limits. August meant fine weather, at least in theory. The US Navy decided to station ships along the route nonetheless, in case any plane needed to put down in the open ocean. It proved a wise move. On 3 August, an oil pump failure forced the *Boston* down. The *Chicago* made contact with the USS *Richmond*, a Navy destroyer, dropping a note with the troubled aircraft's position – tied to the *Chicago's* only life jacket. It was while being towed by the *Richmond*, which by now carried *Boston's* pilots, that the plane capsized and sank.

## HOMECOMING

The flight from Iceland to Ivigtut, in Greenland, was yet another trial of the pilots' skill and courage. Heavy fog meant they had to fly low, close to the waves. Travelling at 145km/h (90mph) and with little visibility, on several

## WORLD CRUISER CHICAGO VITAL STATISTICS

Crew: 2
Length: 10.8m (35ft 6in)
Height: 4.14m (13ft 7in)
Wingspan: 15m (50ft)
Powerplant: 1 Liberty V12 engine, 420HP
Maximum speed: 166km/h (103mph)

© PF-(SDASM4) / ALAMY STOCK PHOTO

Left: America's Douglas World Cruisers, the first planes to complete an around-the-world flight.

# FASTER, HIGHER, LONGER

From 175 days and 74 stops to 67 hours on one tank of fuel, aerial circumnavigations have tested pilots and their planes.

© BETTMANN / GETTY IMAGES

**1924**

First aerial circumnavigation, **US Army Air Service** (World Flight), 175 days, over 42,398km (26,345 miles).

**1933**

**Wiley Post** makes first solo circumnavigation in seven days, 19 hours and 49 minutes (covering 25,105km/15,600 miles).

**Howard Hughes** breaks Post's record, completing his circumnavigation in three days and 19 hours.

© KEYSTONE-FRANCE / GETTY IMAGES

**1938**

American **Geraldine Mock** becomes the first woman to complete a solo aerial circumnavigation, piloting the *Spirit of Columbus* Cessna 180.

**1941**

First circumnavigation by a commercial airliner, **Pan Am's Pacific Clipper.** It flies westward from San Francisco, vai Auckland, to New York, covering 50,700km (31,500 miles) in eight days and 19 hours.

**1949**

US **Air Force B-50 Superfortress** makes first non-stop circumnavigation, in 94 hours and one minute, with four in-air refuellings during the 37,742km (23,452-mile) flight.

**1964**

© RON BURTON / GETTY IMAGES

© H. ARMSTRONG ROBERTS/CLASSICSTOCK / GETTY IMAGES

The first solo circumnavigation by helicopter is made by the Australian philanthropist, **Dick Smith.**

© FAIRFAX MEDIA ARCHIVES / GETTY IMAGES

**Steve Fossett's** *GlobalFlyer* records the first non-stop, non-refuelled, solo circumnavigation in an aeroplane, in a time of 67 hours; Fossett had clocked up 37,000km (23,000 miles) (see p.96).

**1982**

**2005**

occasions they only just missed looming icebergs. From Ivigtut, the crews of the *Chicago* and *New Orleans* made the 900km (560-mile) flight across the Atlantic to the wonderfully named Icy Tickle, in Labrador, Canada. They had flown over the Pacific, Indian and Atlantic Oceans, and endured climatic extremes from Arctic to tropical. While four had departed, only two of the original aircraft completed the full circumnavigation: *Chicago*, flown by Lieutenants Lowell Smith and Leslie Arnold; and *New Orleans*, piloted by Lieutenants Erik Nelson and John Harding Jr.

Once in the United States, the World Flight faced adoring crowds, eager to see America's latest aviation heroes. The planes flew down the East Coast, to Washington, DC, west across the Alleghenies to Dayton and Chicago, and south to Dallas. From there, they crossed the desert, southwest to San Diego. Their triumphant journey up the West Coast culminated in the official conclusion of the World Flight, in Seattle, on 28 September 1924.

## PUBLICITY PRESSURE

Public interest soon turned from expectant to exhausting, and then suffocating. Forced to endure endless rounds of parades, speeches and banquets in their honour, the pilots must have longed to return to the air. Calvin Coolidge seemed to acknowledge as much in his official presidential telegram. 'If our hospitality seems ferocious,' Coolidge wrote, 'forgive us because it comes from the heart. You will find, as you proceed along the home stretch, that these receptions are the first evidence of the feeling that all Americans long to show you. The world never forgets its pathfinders. Those who tread the wilderness and cross the seas filled with dangers, are never forgotten by posterity.'

Below: crowds gather at California's Santa Monica airfield in 1924 to see the the Douglas Cruisers prior to their historic round-the-world flight.

# WORLD BEATERS

The changing shape of long-distance flight.

- SPEED
- RANGE
- POWER
- NUMBER OF PASSENGERS

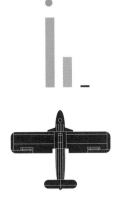

### DORNIER
### DO J
### (1922)

Top speed: 185km/h
Range: 800km
Capacity: 3 crew,
8 passengers

### DOUGLAS WORLD
### CRUISER
### (1924)

Top speed: 166km/h
Range: 3540km
Capacity: 2 crew

### BOEING 314
### CLIPPER
### (1938)

Top speed: 340km/h
Range: 5896km
Capacity: 11 crew, 74
passengers

### BOEING B-50
### SUPERFORTRESS
### (1947)

Top speed:
634km/h
Range: 12,472km
Capacity: 8 crew

### CONCORDE
### (1969)

Top speed: 2179km/h
Range: 7222km
Capacity: 3 crew,
120 passengers

### AIRBUS
### A380
### (2007)

Top speed: 945km/h
Range: 14,800km
Capacity: 2 crew,
575 passengers

# AMELIA EARHART: THE LOST ACE

Arguably the most famous aviator of them all tragically – and mysteriously – failed in her 1937 attempt to become the first woman to fly around the planet. But by then Amelia Earhart's guts, talent and style had made her not just America's sweetheart but one of the first global superstars.

Amelia Earhart was destined to be an icon. In 1934, at the age of 37, she was named one of the US' best-dressed women by the Fashion Designers of America. Fast forward 82 years, and Earhart was a headline act in *Goodnight Stories for Rebel Girls*, the bestselling children's anthology of female pioneers and feminists published in 2016, alongside the likes of Cleopatra and Catherine the Great.

Earhart's best-dressed award was a rare dalliance with an industry she regarded as trivial, at least compared with the occupation which was to make her name. From the beginning, Earhart set out to defy convention. Born in July 1897, beneath the vast skies of Kansas, she was, even as a teenager, an avowed petrolhead; she took courses in car mechanics and worked as a delivery driver. Earhart claimed to have known aviation was her true destiny from the very first time she flew.

## THE FLYING BUG

On 28 December 1920, when she was 23, Earhart accompanied her father to a California air show. Edwin Earhart, a successful lawyer for the railroad industry, was only too happy to pay a few dollars for his daughter to enjoy a short flight around the airfield. By the time that small plane was a few hundred feet off the ground, Earhart had decided to make flying her career.

From that moment, Amelia Earhart was a woman in a hurry upward. She took flying lessons and the following year, 1921, bought her first plane, a yellow Kinner Airster biplane she nicknamed the *Canary*. Just 12 months later, Earhart set a new altitude record for a female pilot, 4267m (14,000ft); by 1923 she had gained a full pilot's licence, one of only a handful of women in the US to hold one.

In 1928, eight years after that first flying experience, Earhart was invited to join a mission that would make her the first woman to fly across the Atlantic. Acting as its navigator, Earhart boarded a Fokker F.VII in Newfoundland on 18 June and, some 20 hours and 40 minutes later, pilot Wilmer Stultz touched down in Burry Port, South Wales. 'Most of the way I was flying blind because of the fog and rain', Stultz told the local *Llanelli Mercury*. 'We landed here in South Wales because we were short of fuel.' Such was the national press interest in the flight, reporters who had travelled from London struggled to find a free telephone line.

That newspapers went crazy for Earhart was hardly surprising. From Delhi to the Dust Bowl, Earhart became one of the most famous people in the world. In her quest to break boundaries, she was assisted by Amy Johnson, who, in 1930, became the first woman to fly solo from England to Australia. On her own subsequent tour of that continent, Johnson was also mobbed wherever she went.

In 1931, Earhart married George Putnam, a publicist who would go on to deploy his considerable influence and contacts to promote his wife's aerial excursions. Earhart was forthright, demanding if not a prenup, then undertakings from her fiancé that he would not stymie her ambitions. 'Please let us not interfere with the other's work or play', Earhart wrote to her betrothed on the morning of her wedding day. 'I cannot guarantee to endure at all times the confinement of even an attractive cage'.

So keen was Earhart to liberate others from an 'attractive cage' that, in 1929, she had founded the Ninety-Nines, a women-only pilots association that to this day provides advice

Right: Amelia Earhart lands in Northern Ireland after her groundbreaking transatlantic flight of 1932.

and mentoring for female pilots. Named for the 99 licensed women pilots who attended the inaugural meeting, of a possible 117, today membership of the Ninety-Nines runs to more than 5000 aviators.

## ACROSS THE ATLANTIC

In May 1932 Earhart was at it again, becoming the first woman to fly solo across the Atlantic. Taking off from Newfoundland in her Lockheed Vega 5B, and with an estimated 10 hours of flying time remaining, problems beset the flight early on. A fire on board was followed by Earhart's realisation that a fuel-tank gauge was dropping faster than was normal. By now four hours into the flight, she pressed on, eventually landing in a farmer's field in Culmore, in Northern Ireland. The famous exchange that took place when Earhart was greeted by a farmhand was reported by *Life* magazine. 'Have you flown far?' Earhart was asked. 'From America', came the reply. Her total flight time for the 3220km (2000-mile) trip had been 14 hours and 56 minutes.

Earhart was welcomed back to New York with a ticker-tape parade and later entertained in Washington as the personal guest of former President Calvin Coolidge. She was also

### THAT'S MISS EARHART TO YOU

When the *New York Times* had the temerity to congratulate Amelia Earhart by her married name – 'Mrs Putnam flies Atlantic in Record Time', ran the headline – she wrote to both editor and publisher personally to insist they not make the same mistake again.

awarded the Distinguished Flying Cross, the first woman ever to receive it.

The flight which was to cement her legend was also the one that would seal her fate.

In 1935, Earhart had become the first woman to complete a solo flight from Hawaii to San Francisco, a distance of 3860km (2400 miles). After proving she could cross the Atlantic and much of the world's largest ocean, plans were made for attempting another record – the first circumnavigation of the world by a female pilot. The first aerial circumnavigation of any kind was made in 1924, by the crews of the World Flight (see p.76); the first solo navigation was made in 1933, by America's Wiley Post.

Aware of the huge costs involved in such an undertaking, Earhart set about fundraising, establishing, in 1936, the Amelia Earhart Fund for Aeronautical Research. Her flight companion and navigator would be Fred Noonan, a vastly experienced sailor but also the pilot who had plotted many of the first commercial airline routes across the Pacific; and who, crucially, was renowned for navigating by the stars. When the research fund hit $80,000, the pair purchased a Lockheed Electra 10E, a twin-engine all-metal monoplane they planned to pilot around the entire planet. A decade and a half previously, when Earhart first took to the California skies, she had been aboard a biplane whose range would have been, at most, 240km (150 miles).

Together, Earhart and Noonan made two attempts at a continuous world flight. With a plan to fly west, at approximately equatorial latitudes, the first took off from Oakland, California, on 17 March 1937. On reaching Hawaii, however, Earhart and Noonan were forced to abort; during take-off for their second leg, their landing gear collapsed causing extensive damage. Repairs were sufficiently long-running that Earhart and Putnam were required to undertake additional fundraising. On 21 May 1937, they climbed into the skies above Oakland for a second time – on this occasion, for reasons of winds and weather, heading east. Only once they reached Miami did Earhart reveal to the press that this was, indeed, a second attempt at realising her round-the-world ambition. The two aviators flew next to South America, then on to Africa, India and Southeast Asia, proceeding at what Earhart reported as a 'leisurely' pace.

## 'FUEL IS RUNNING LOW'

On 2 July 1937, around mid-morning, Earhart and Noonan launched their Lockheed Electra along a runway on the island of Lae, in New Guinea. As its wheels lifted, they had covered

Below: Amelia Earhart became an instant celebrity following her 1928 transatlantic flight. Left: Earhart's Lockheed Electra soars over the Golden Gate Bridge in 1937, at the start of the ill-fated circumnavigation attempt.

more than 35,400km of the 46,670km (22,000 of the 29,000 miles) separating their start and finish point. 'Please know that I am aware of the hazards', Earhart had written to her husband before her second world flight attempt. Among Earhart's team, the leg from Lae to the next stopover, Howland Island, was acknowledged as particularly difficult. More than 4020km (2500 miles) of the Pacific lay between take-off and landing. It was marginally further than Earhart's previous oceanic sorties, across both the Pacific and Atlantic, but a crucial difference was the size of the target; Howland, a remote atoll, was barely 2000m (6500ft) long.

The journey time to Howland was estimated at around 18 hours, flying through the night for a dawn landing, thus enabling Noonan to call upon his skills of celestial navigation. Waiting for them, with fuel, was the *Itasca*, a US Coast Guard cutter. Precisely what happened next remains the subject of ongoing debate. Scanning the airwaves, the *Itasca*'s radio operator picked up a transmission at 7.42am. 'Fuel is running low', Earhart said. 'We are flying at 1000ft.' The signal was strong, the two aviators close, but visual contact could not be established. For the next hour, the *Itasca* radioed the plane but each time received only new, outgoing messages. The last came one hour after the first, at 8.43am: 'Calling *Itasca*,' Earhart's voice said. 'We are on the line 157–337; we are running on the line north and south.' The numbers referred to a position, along a line that ran 157 degrees southeast, and 337 degrees the other way, northwest. Running that line 'north and south' suggests Earhart and Noonan were criss-crossing the area, desperately trying to locate the ship.

That was the last that was heard of them. Despite searches, neither the plane, nor either body, was found. In terms of cause for a probable crash, scattered cloud that morning would, almost certainly, have impaired Noonan's vision. The rest is guesswork. Rumours have been circulating ever since. One held that Earhart had been taken prisoner by the Japanese and brainwashed, alleging that during WWII she broadcast as Tokyo Rose on the Axis-propaganda station which sought to demoralise Allied forces in the Pacific theatre. Another story ran that she was spying for the US Air Force – mission complete, Earhart returned home and disappeared all over again.

In 1940, human remains were found on Nikumaroro Island, several hundred miles southeast of Howland. An investigation proved inconclusive, as did subsequent research, as recently as 2018. In 2019, Robert Ballard, the oceanographer who, in 1985, located the wreck of the *Titanic*, announced his own expedition to track down Amelia Earhart, aviation's lost ace.

# THE HARD WAY ROUND

When the 1941 attack on Pearl Harbor interrupted the return flight of Pan Am's Pacific Clipper, its skipper and passengers buckled up for a very bumpy homeward journey.

On the morning of 8 December 1941, Pan Am's *Pacific Clipper*, a luxurious Boeing 314 flying boat, was preparing to land in Auckland, New Zealand. A scheduled stop on the airline's transpacific service, it was six days from its home port of San Francisco. Pan Am's Clipper fleet were the last word in glamour; a one-way ticket from San Francisco to Hong Kong cost $760, the equivalent of $13,000 today. With a top speed of 338km/h (210mph) and a range of 5930km (3685 miles), Clippers offered fully seated dining rooms serving five-star menus. Refuelling at Pacific islands and atolls, Pan Am built luxury hotels at locations so remote (and beautiful) only Clipper passengers and staff could visit them.

It was in Auckland that the 10-man crew learned of the Japanese attack, the previous day, on the US fleet at Pearl Harbor. All across the Pacific, Pan Am facilities had come under assault: at Wake Island, the *Philippine Clipper* flying boat had hastily boarded staff before making its escape, riddled with bullet holes; in Manila, docked seaplanes were severely damaged; and in Hong Kong, two Pan Am Sikorsky S-42Bs destroyed.

The *Pacific*'s captain, Robert Ford, was faced with a problem: how to get his passengers home? After a week at Auckland's US Consulate General, Ford finally received word from Pan Am HQ – they were to return to the US, flying westward. As for gasoline and supplies, they would be on their own, and flying over unfamiliar terrain. Ford took off on 16 December, bound for New Caledonia to collect stranded Pan Am staff, before heading for Australia. They put down next in Gladstone, north of Brisbane. The next day, Ford and the *Pacific Clipper* headed northwest, to Darwin, flying over the Queensland desert and watching as it gradually transformed into tropical rainforest.

From Darwin they made for Surabaya, in Indonesia, and it was then things got sticky. With no intelligence available, Ford had to assume the Japanese expansion had not reached this far as his crew steered the massive flying boat, whose vast underbelly presented an enticing target for an

enemy ship's gunner, across 2250km (1400 miles) of open ocean. They reached Surabaya, only to be intercepted by a suspicious British fighter squadron. Escorted to relative safety, the *Pacific* was still required to taxi into port, through waters heavily mined against Japanese submarines.

After refuelling with automobile-grade gasoline – none of the *Pacific*'s preferred 100-octane fuel was available – Ford headed for the port city of Trincomalee, in Ceylon (now Sri Lanka). Flying with no charts, but simply the coordinates of their destination, it was down to the skills of their navigator, Roderick Brown, that they found Ceylon. A safe landing followed, but only once the crew was sure a patrolling Japanese submarine had disappeared into the distance. Refuelling once again, the *Pacific* left Trincomalee on Christmas Eve, only to turn back after losing an engine. Christmas Day was spent on repairs, before the *Pacific* finally took to the air once more, on 26 December, headed for Karachi.

After a blissfully uneventful flight, Ford continued on to Bahrain, and then across the vast desert expanse of the Arabian Peninsula, to Khartoum in Sudan, where they landed on the River Nile. Not wishing to risk any further desert flying – the *Pacific*'s propellers had struggled in the sand-rich air – the crew pressed on to

Below, from top: in the wake of Pearl Harbor, Captain Robert Ford's impromptu round-the-world flight returned the *Pacific Clipper*'s passengers safely back to the US; with its seated dining rooms and lavish service, the *Pacific Clipper* flying boat was the last word in glamorous travel.

*Fly to* **HAWAII** *by Clipper*

**PAN AMERICAN WORLD AIRWAYS**
*The System of the Flying Clippers*

Léopoldville, in the Belgian Congo (now Kinshasa in the Democratic Republic of the Congo). From the air, the broad Congo River appeared a welcoming landing strip.

The reality was different – the atmosphere oppressively hot and the river given to strong currents. When it took off next, the flying boat clawed its way into the sky, doing so just before it reached a set of perilous rapids. Safely clear of obstacles, it droned westward for 5766km (3583 miles) to Natal in Brazil, then up the coast to Port of Spain, Trinidad. Finally, on 6 January 1942, the *Pacific Clipper* reached the marine terminal at LaGuardia Airport in New York. Since leaving San Francisco, the *Pacific Clipper* had logged a total flight time of 209 hours and 50,700km (31,500 miles). If not quite a true circumnavigation, it was the longest flight by a commercial airliner – and they had certainly done it the hard way.

Above: early advertising for Pan Am's Clipper fleet – first stop Honolulu.

## ... AND THE EASY WAY: COCKTAIL BEFORE DINNER, SIR?

One of the more striking features of the first scheduled passenger service around the world was that it came so soon after the end of WWII. In 1947, barely 24 months since the skies over Europe's cities had been filled with enemy planes and anti-aircraft fire, Pan Am launched flight 001, the first service of its kind.

Given the state of European infrastructure following the war, it was hardly surprising that it was a US company that made this pioneering move. Pan Am's westbound service departed from San Francisco and made nine scheduled stops before returning whence it came. The first stop, Honolulu, was in the now conflict-free Pacific, the second in Hong Kong. Thereafter, the flight put down in Bangkok, Delhi, Beirut and Istanbul before landing in occupied Germany, in Frankfurt. London and New York completed the circumnavigation of a service that was designed very much as hop-on/hop-off.

In fact, provided the 10 legs were completed within six months of purchasing the original ticket, it was possible to stay in any of the destinations for as long as you wished; those wanting to complete the flight in one fell swoop could expect to be back in San Francisco in around two days. Which was a good way to burn through cash – in 1947, an economy-class ticket cost $2000. (The average US family income in the bottom fifth of earners at the time was around $7800 per annum.)

As the short-term austerity imposed by the war receded, the emphasis was very much on luxury; by the mid-1950s, Pan Am was marketing its round-the-world flights as cocktail parties in the sky, a strategy it had successfully developed with its pre-war Clipper services. An eastbound version, Pan Am flight 002, launched that same inaugural year, offering an identical service in the opposite direction.

# 'IF I DON'T GET OUT OF THIS HOUSE, I'LL GO NUTS'

**Jerrie Mock overcame war, a competitor and endless condescension to become the first woman to fly solo around the world in 1963.**

At the age of 11, in 1936, Geraldine Mock (née Fredritz) was given an atlas. A children's edition, it featured, she recalled years later, jungles and elephants; she also remembered the Pyramids of Egypt and Victoria Falls, in what was then Rhodesia. That she resolved there and then to travel the world was not unusual; the fact that she eventually did so, three decades later, having settled down and raised a family, certainly was. After all, her native Midwest America, in the years following the war, was a byword for conformity and convention.

A product of her time, 'Jerrie' Mock became known, with unerring predictability, as 'the flying housewife'. In her twenties, she enrolled for an aeronautical engineering degree at Ohio State, but sacrificed any potential career before it had even started, relinquishing her studies once she met her future husband, Russell.

The three children they raised might have stymied less formidable women, yet the family home also acted as a spur; for one thing, there was a healthy rivalry between husband and wife. In 1954, at the age of 29, Mock took flying lessons – Russ had come home from work and announced he was starting lessons that coming weekend, a Saturday. Mock acquired a licence on the preceding Friday and took her first lesson the same day.

Together, the Mocks owned a succession of small planes and, eventually, a Cessna 180, named *Spirit of Columbus*. It was the plane that would take Jerrie Mock around the world, and Mock confessed she'd bought it because its name was similar to the *Spirit of St Louis*, the plane Charles Lindbergh had piloted across the Atlantic in 1927.

In the daredevil, masculine world of aviation, the Mocks were a homely novelty. They had the air of picnickers, taking to the skies with a few

sandwiches, a flask and a blanket. They flew to Canada for one vacation, to Bermuda for another. Jerrie flew in air races, 'always coming last'. Nevertheless, by the time Mock took off on her solo circumnavigation from Columbus, Ohio, on 19 March 1964, she had managed 750 hours of flight time. (In the UK today, aspiring aviators need around 60 hours to get a basic licence; for commercial flyers that increases to 250 hours to fly as a co-pilot.)

In late 1963, when Mock planned to fly across the Atlantic, it was her husband who suggested that, should she make it, she might want to keep going; her 'world flight' was born. (The story goes that the seed of the idea was planted when an exasperated Mock declared to her husband 'If I don't get out of this house I'll go nuts.') Mock was wary of revealing her plans to friends, not simply because of the social mores that said a woman's place was at home, but because she was aware no woman had yet made a successful solo flight around the world (Mock's idol was Amelia Earhart). So, she kept to herself the idea that 'plain old Jerrie Mock' might be the one to complete what Earhart had started.

## AN 'UNLADYLIKE' PURSUIT

Mock's mother deemed flying 'unladylike', yet the daughter found surprising allies. For example, her flight path was plotted by Colonel Lee Lasseter, a distinguished Marine Corps pilot who, in 1972, became the only marine aviator to shoot down a North Vietnamese MiG fighter during the entire conflict. On Lasseter's advice, Mock also wrote – assiduously, as it turned out – to aviation officials in each of the countries she planned to visit. Learning by heart the local rules and regulations, in bed at night she would rehearse the various legs of her flight.

Travelling east, Mock's 37,000km (23,000-mile) journey saw her cross the Atlantic, Europe and the Middle East, and then on to Southeast Asia and the Pacific. Flying over Vietnam – it

# BY PLANE

Clockwise from left: Geraldine 'Jerrie' Mock and her Cessna, *Spirit of Columbus*, before departure; a floral welcome in Columbus, Ohio, after completing the circumnavigation; President Lyndon Johnson presents a FAA Gold Medal to her in 1964.

would be another year before the first US ground troops were deployed in the region but the wider conflict was already a decade old – Jerrie was struck by how peaceful it looked.

Jerrie Mock landed the *Spirit of Columbus* back in Ohio on 17 April, clocking a total journey time of 29 days, 11 hours and 59 minutes. Though Mock received a special commemorative medal from President Lyndon Johnson and was later awarded the Louise Blériot medal by the Féderation Aéronatuique Internationale, she seemed content to return to her quiet life. At the time of her death, in 2014, Mock had yet to be inducted into the US National Aviation Hall of Fame. Indeed, only as the gender gap was finally redressed over subsequent decades did later generations celebrate her achievements.

An interesting subplot to Mock's story exists in the shape of Joan Merriam Smith, an experienced pilot who made her own bid to be the first woman to fly around the world, taking

off two days before Mock, on 17 March 1964. The date, precisely 27 years after Earhart began her own ill-fated endeavour, from Oakland in California, was symbolic; Smith took to the skies from the same Oakland airport.

The press did their best to paint them as rivals, the homemakers in the sky (when Mock returned she was asked, during a rare television appearance, who had cleaned her house while she was away). If Mock feared being beaten to the record, she did not reveal it, although one imagines at least a small sigh of relief when mechanical problems forced Smith to abandon her bid after just a few days.

So it was Jerrie Mock, 'the flying housewife', who got there first. And more important to her than any medal was the telegram she received after completing her circumnavigation of the planet. One among thousands, it came from Amelia Earhart's sister, Muriel. It read: 'I rejoice with you.'

# SLIPPING THE BONDS OF EARTH — THE LONGEST HAUL OF ALL

Super-efficient jet engines, vast fuel tanks, solar power... aviators have reached for every revolutionary high-tech advantage to achieve the same goal: a non-stop circumnavigation.

The three decades spanning the late 20th and early 21st centuries have seen aerial circumnavigators and their engineering teams embrace a a huge range of technology in pursuit of the same simple goal: to fly around the world non-stop.

## VOYAGER (1986)

**The first non-stop flight around the world**

Designed by veteran aerospace engineer Burt Rutan, the *Voyager* aircraft had a sole purpose – to fly non-stop around the world. That meant compromises. *Voyager* was not so much a plane as a flying fuel tank; 17 tanks in total, the largest of which were the three fuel cells located in each wing.

In Rutan's carbon-fibre design, the wings were also extremely – the untrained observer might say unnaturally – flexible. Removing wing struts saved weight. Once airborne, lift would then support both wing and aircraft. On the ground, however, when bounced up and down, thousands of litres of highly flammable aviation fuel could be heard sloshing around inside each wing.

*Voyager's* wings also featured a 'canard' construction, a small forewing located in front of each main wing and designed to help prevent the aircraft stalling. Unsurprisingly, much of the design involved trial and error. For example, the metal propellers that were originally discounted as too heavy were later found to be significantly more efficient than wooden alternatives, thus contributing many more kilometres per litre.

In contrast to the dynamic pioneers of modern aviation research, the Rutans were a pragmatic, almost hangdog bunch. *Voyager's* pilots were to be Burt Rutan's younger brother, Dick, and Jeana Yeager, both experienced aviators. 'I got to really hate this airplane,' Dick later told a press conference. 'I felt not only was it not going to work but I would probably die in it.'

*Voyager's* cabin was unpressurised. The flight plan was to maintain an altitude of around 2440m (8000ft), then use bottled oxygen if they needed to fly higher, where tailwinds, likely to prove important, might be found; with oxygen supplies limited, the pilots would have to carefully judge when best to use it. There was also the issue of bad weather. Indeed, the mission's senior meteorologist, Len Snellman, confessed that when flying in the Intertropical Convergence Zone (ITCZ) – the 'trade winds' of the equator – sufficient turbulence might destroy the plane completely.

Burt Rutan was sanguine about the physical toll the flight might take on his brother. 'Dick would be the last one to talk about it,' he told a television documentary ahead of the flight, 'but he's not 20 years old any more... little things creep up on you.'

To save yet more weight, food was strictly rationed (water was not); Yeager also cut her hair. Finally, after months of preparation, on the morning of 14 December 1986, Rutan and Yeager buckled up at Edwards Airforce Base in California. As *Voyager* approached the 100 knots required for take-off, the fuel-laden wings

*Voyager* pilots Dick Rutan and Jeana Yeager.

94

© BETTMANN / GETTY IMAGES

proved so flexible their tips scraped the runway. In the event, *Voyager* rolled for more than a mile of Edwards' three-mile airstrip before the wings generated sufficient lift.

Flying west to east, it was over the Indian Ocean that *Voyager* surpassed 20,168km (12,532 miles), the previous record for non-stop flight, set by a US B-52 in 1962. The bad weather Rutan feared over central Africa did not materialise and they headed next for Brazil, and finally home.

A large crowd greeted *Voyager*'s landing and the pilots emerged, exhausted and barely able to stand. Rutan, who lost 2.7kg (6lb) in weight, and Yeager, who lost 4kg (9lb), claimed they had slept for no more than a few hours. They had eaten just 10% of their food supplies. The flight distance was officially recorded as 40,212km (25,000 miles), in a time of nine days, three minutes and 44 seconds.

### VIRGIN ATLANTIC GLOBALFLYER (2005)

**The first solo non-stop, non-refuelled flight around the world**

In March 2005, when the record-chasing billionaire Steve Fossett announced plans for the first solo non-stop flight around the world, his target was so ambitious it seemed almost arrogant. *Voyager* had taken 216 hours

### JUMPING FOR THEIR LIVES

A dozen years after the *Voyager* flight in 1986, Dick Rutan attempted another unprecedented non-stop circumnavigation, by balloon. But after just three hours' flight, he and co-pilot David Melton were forced to parachute to safety over New Mexico (a helium cell burst). Another team took the record the following year (see p.188).

– Fossett determined to do it in less than 80. What set his Virgin Atlantic *GlobalFlyer* aircraft apart from its predecessor was, in part, two decades of aviation design. But it was also down to a whole heap of money. The cost of *GlobalFlyer*'s construction, undisclosed to this day, was likely in the region of several million dollars. That was a drop in the ocean – as the state-of-the art plane's name reveals, Fossett's partner on the project was fellow billionaire, Richard Branson.

There were similarities though. *GlobalFlyer*'s designer was none other than Burt Rutan, the aerospace engineering legend who, in addition to creating *Voyager*, was also behind the world's first privately funded spacecraft. Improved resources brought enhanced materials. *GlobalFlyer* was built from graphite-epoxy, a lighter, stronger form of the carbon-fibre used in *Voyager*.

A crucial difference was the power source. *Voyager* corkscrewed its way through the air using propellers; *GlobalFlyer* was driven by a high-efficiency jet engine. It was, in fact, a high-tech hybrid: part high-altitude jet, part glider. If successful, Fossett, who already held the circumnavigation record for a balloon (see p.196), would have to spend his planned 80

© POOL / GETTY IMAGES

hours in a cockpit 2m (6.5ft) long; positively luxurious, compared with *Voyager*'s bathtub-sized entertaining area. The advances in design were notable even during take-off. Two decades previously, *Voyager* had wobbled down a California runway like a drunken albatross. On 1 March 2005, *GlobalFlyer* powered along the tarmac at Kansas' Salina Airport with the bristling intent of a military fighter.

Flying west with the jet stream, *GlobalFlyer*'s power-to-weight ratio was designed to increase in Fossett's favour. Just as it had for *Voyager*, the vast majority of *GlobalFlyer*'s weight sat in its fuel tanks; on departure, this was as much as 85%. Yet as Fossett burned fuel, the already optimised jet engine would only become more efficient, meaning that the further the plane flew, the faster it would get. And then some. A pressurised cabin enabled Fossett to fly at an altitude of around 14,000m (46,000ft), and at an average speed of 480km/h (300mph). That's 3050m (10,000ft) higher than commercial airline routes and a staggering 11,580m (38,000ft) higher than its non-stop predecessor, *Voyager*.

The mission wasn't entirely without incident. *GlobalFlyer* inexplicably lost 1200kg (2600lb) of fuel early in the flight and Fossett considered aborting as he reached Hawaii. Only when he picked up a strong tailwind did he decide to press on. Given the technological advantages, however, it is hardly surprising Fossett smashed even his own target, touching down on the Salina blacktop, on 3 March 2005, a total of 67 hours and one minute after setting off.

Left: the *Voyager* in flight over the Mojave desert. Below, from left: Virgin Atlantic's *GlobalFlyer* above the Atlas Mountains during its non-stop circumnavigation in 2005; *GlobalFlyer* pilot Steve Fossett with Richard Branson.

## SOLAR IMPULSE 2 (2015–16)

**The first flight around the world without fuel**

This near-silent aircraft made a big noise when it took off from the United Arab Emirates on 9 March 2015. Sixteen and a half months later, *Solar Impulse 2* completed a circumnavigation of the Earth using nothing but solar power. In doing so, it just might have redefined the nature of aviation. Conceived by the Swiss businessman and entrepreneur André Borschberg, the aim of the privately funded project was to explore the potential of clean, namely long-range solar, aviation technology. A prototype, *Solar Impulse 1*, completed a successful test flight in December 2009.

*Impulse 2*'s promotional videos showed a team and crew for whom glasses were clearly three halves full. And why wouldn't they be, when sponsors included sovereign wealth funds from the United Arab Emirates and Google? .

The craft itself isn't so much high-tech as science fiction. Viewed from the ground, the vast illuminated wingspan makes *Impulse* resemble a spacecraft coming in to land. In one marketing video, rapid-fire news bulletins from various foreign-language broadcasters urgently presaged something momentous. Still, it's hard to deny an aircraft capable of flying around the clock without fuel might just be a game-changer. (Its wider remit was to promote clean aviation technology, a timely intervention given some estimates suggest global aviation could consume 5% of the Earth's remaining carbon budget by 2050.)

© EDMOND TERAKOPIAN - PA IMAGES / GETTY IMAGES

Solar Impulse 2's circumnavigation was billed as much as a tour as a record-breaking flight, with India its first destination. A large press corps awaited touchdowns, and the pilots – Borschberg flew alternate legs with Bertrand Piccard (who previously co-piloted *Breitling Orbiter 3* to complete the first non-stop balloon flight around the world, see p.190) – inevitably emerged to rapturous applause.

Unlike *GlobalFlyer*, *Solar Impulse 2* was fragile, constructed almost entirely from solar panels. Aesthetically, it looked more like a Wright brothers' design than its jet-powered predecessor. And it was slow. Flying from Nagoya, Japan, to Hawaii, a distance of 7211km (4481 miles), took Borschberg 117 hours and 52 minutes, when the standard flight time for a

Above: *Solar Impulse 2* glides over Abu Dhabi, UAE, at the start of its 2015 circumnavigation. Below: André Borschberg in the pilot seat of *Impulse 2*.

commercial airliner is eight hours. In five days and nights, Borschberg broke seven aviation world records, including the world's longest flight by a solar-powered aircraft (separate records for time and distance) and also the longest solo flight in terms of time. Success came at a price. The long, long haul to Hawaii caused the planes batteries to severely overheat, the result of too much insulation, and consequently *Solar Impulse 2* was put into an enforced nine-month 'hibernation'; wrapped in foil blankets, to give the panels some rest.

After this unexpected hiatus, the journey resumed on 26 April 2016, with Piccard back at the helm. Hype continued to build. During the Hawaii–San Francisco leg of *Impulse's* tour the Paris Climate Agreement was being finalised in France; Piccard even addressed the conference from *Impulse's* cockpit. When Borschberg returned, briefly, to *Impulse's* base in Switzerland after recovering from shingles, he was greeted like a movie star. And on landing in New York after crossing North America, he was welcomed personally by the then UN Secretary General, Ban Ki-moon.

The transatlantic crossing took *Solar Impulse 2* to Seville, then on to Africa, where it soared over the Pyramids before touching down in Cairo. It returned to Abu Dhabi, touching down silently on 26 July 2016.

© HANDOUT / GETTY IMAGES

## AIR MILES

The five longest non-stop
commercial passenger flights are:

**1** New York–Sydney, 16,200km,
c19h (Qantas, in testing)
**2** Singapore–New York, 15,343km,
c18h 30m (Singapore Airlines)
**3** Auckland–Doha, 14,535km,
c17h 45m (Qatar Airways)
**4** Perth–London, 14,499km,
c17h 30m (Qantas)
**5** Dubai–Auckland, 14,201km,
c17h 15m (Emirates)

Technological developments in
materials science, wing and engine
design, and more efficient fuels
are contributing to these feats.
However, in the global aspiration to
achieve 'net zero' carbon emissions
in the coming decades (aviation
is responsible for 2.5% of global
emissions), these advances may not
be enough to avoid huge questions
facing the airline industry.

SOLAR IMPULSE 2      BOEING 747-400

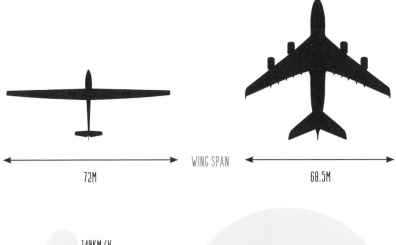

WING SPAN

72M      68.5M

140KM/H

MAXIMUM
SPEED

# 988KM/H

# 5-6 DAYS

MAXIMUM
FLIGHT
DURATION    17 HOURS

8183KM    MAXIMUM RANGE    13,805KM

PASSENGERS

1

$CO_2$ EMISSIONS
PER HOUR (KG)

# 524

0

92

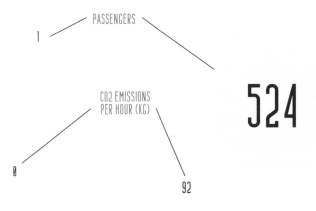

# BY BOAT

When British adventurer Francis Chichester set sail in 1966 to make a solo circumnavigation of the world by boat, he was following in the wake of some big fish. Despite limited sailing experience, it was Chichester's ambition to match, or even beat, the great wool clippers (ships, not shears) that plied their trade between Britain and Australia in the second half of the 19th century. Not only that, he aimed to be the first to complete the voyage single-handed. (Seven decades earlier, Joshua Slocum completed the first-ever solo circumnavigation by boat, but not via the clipper route.)

The captains of the fastest clippers prided themselves on their speed, bringing wool from Australia (and later tea from China), in around 100 days. What made such achievements more impressive was their route, the infamous 'eastabout' passage, through the monstrous seas of the southern oceans. It was a route Chichester planned to follow closely, and which would take him past the three great capes of the southern hemisphere: the Cape of Good Hope, in South Africa; Cape Leeuwin, the most southwesterly point of Australia; and South America's dreaded Cape Horn.

Chichester was an old-fashioned hero in a rapidly modernising age, a conservative at sea as the counterculture thrived at home. He inspired a generation of stalwart sailors, man-against-the-elements imitators who proved every bit as intrepid; the likes of Robin Knox-Johnston and, later, Tony Bullimore, who was rescued after capsizing thousands of miles from help. The ill-fated Donald Crowhurst, on the other hand, was sadly deluded, setting sail aboard a vessel he couldn't handle in a race he couldn't hope to win.

Among the most accomplished sailors to rise to Chichester's single-handed challenge were women such as Poland's Krystyna Chojnowska-Liskiewicz, who in 1976 became the first woman to sail around the world. She was followed by a veritable armada of daring circumnavigators, such as Tracy Edwards, who shattered clichés, or Ellen MacArthur and Dee Caffari, who went ever faster and in both directions.

The high seas show no signs of losing their appeal. For instance, Dutch teenager Laura Dekker took to her yacht despite a ruling by the courts that she was not old enough to sail alone. In 2019, the Australians Riley Whitelum and Elayna Carausu notched their sixth year of an ongoing, crowdfunded circumnavigation. To celebrate, they boosted their 1.15 million online followers by offering a transatlantic lift to climate activist Greta Thunberg.

# THE ADVENTURER WHO LAUNCHED THE SECOND ELIZABETHAN AGE

A 65-year-old who had only taken up sailing a decade earlier. A lung cancer patient who had been given six months to live. A solo sailor who hated his boat. Francis Chichester left Plymouth Sound in 1966 to sail around the world as an unlikely prospect. He returned 226 days later a wildly celebrated hero.

F rancis Chichester's plan, to be the first to sail around the world single-handed, was a bold one given that, in 1958, he had been diagnosed with terminal lung cancer. The prognosis hadn't been good; he was given just six months to live but, eight years later, there he was. By 1966, as he prepared *Gipsy Moth IV*, the 16m (53ft) ketch that would eventually become a maritime legend, Chichester's poor health was not the only remarkable thing about him. Then aged 64, he had only learned to sail in his fifties; allowing for numerous interruptions for spells of illness, Chichester was hardly a salty old sea dog.

But he had been a good many other things: a lumberjack, an estate agent (briefly) and the director of a map-making company. During WWII, Chichester had worked as a cartographer, although it was hardly the wartime role he craved. His three attempts to join the Royal Air Force, as a fighter pilot, came to nothing, due to the fact that he lacked the required 20/20 vision.

This remained a sore point for the Devonian, not least because he was already a qualified pilot and, by his own estimation, a good one. (*Gipsy Moth* took her name from Chichester's beloved De Havilland aircraft.) In 1931, he became the first to fly solo across the Tasman Sea, from New Zealand to Australia. The account he subsequently published, *Alone Over the Tasman Sea*, became a classic of aviation literature.

## AN UNLOVED LEGEND

Among sailors, *Gipsy Moth IV* is as iconic a vessel as you'll find, her historical importance sufficient enough to forgive her many quirks. Chichester's own view of the boat was pretty much the opposite, however. The fourth of five boats that Chichester commissioned, *Gipsy Moth IV* was built in Hampshire, at the Camper & Nicholsons yard in Gosport. The company can trace its boat-building history back to the 18th century, although today it specialises in luxury yachts.

From the moment *Gipsy Moth IV* launched, Chichester hated the boat. His principal objection was her size. While he had maintained contact with the designers after the plans were drawn up, Chichester felt that the finished boat was too big. The designers in question, John Illingworth and Angus Primrose, were known for their racing yachts and were well regarded, their designs having won countless races. Chichester thought his brief had been clear enough: design a boat to carry one man around the world, and quickly. In the event, Illingworth and Primrose pointed to the compromises they had needed to make.

*Gipsy Moth IV* was significantly over budget, too, coming in at almost twice the estimated price of £12,500 ($16,000). And while Chichester was a practical man, he was not one inclined to ignore bad nautical omens. When *Gipsy Moth IV* was at last ready for launch, the official ceremony was beset by problems; the boat got stuck on the slipway and the bottle of champagne due to be sacrificed for her christening refused to smash open.

Right: Francis Chichester aboard *Gipsy Moth IV*, 1967.

With trials taking place on the Solent almost immediately, *Gipsy Moth IV* did little to endear herself to her new master. Created from layers of laminated mahogany, the fabrication of the boat would, today, be regarded as old-fashioned, certainly compared with the Kevlar and carbon-fibre builds of modern ocean-going catamarans.

One of Chichester's principal beefs rested on the boat's performance. To his mind, *Gipsy Moth* felt tentative 'heeling', that is, when leaning under sail. Chichester also complained about the deck, which he found slippery, so much so that during the Solent trials he fell and injured his leg – an incident that only made him dislike the boat more. Lastly, he noted that the distribution of the sails was likely to be problematic. In its standard rigging, *Gipsy Moth* was designed to support 79 sq metres (854 sq ft) of sails. These had to be handled by one man, and an ageing, lung-cancer patient at that.

## THE ADVENTURE BEGINS

Chichester set sail from Plymouth on 27 August 1966, commencing his voyage at the harbour breakwater. His stated aim was to match the speed of the great clippers, the cargo ships that, in the later 19th century, carried valuable wool from the homesteads of Australia to the factories

## FEAR OF THE HORN

'It would be fair to say [Cape Horn] terrified me ... I told myself for a long time that anyone who tried to round the Horn in a small yacht must be crazy. Of the eight yachts I knew to have attempted it, six had been capsized or somersaulted before, during or after the passage.' Francis Chichester, *Gipsy Moth Circles the World* (1967)

of Britain. A decent clipper might expect to complete the voyage in 100 days, but that meant following the 'eastabout' passage, the infamous route through the southern oceans that took in the southern hemisphere's three great capes: Good Hope, in South Africa; Leeuwin, the most southwesterly point of Australia; and South America's dreaded Cape Horn.

Though Chichester's reservations about the sail configuration were valid, they were partly a problem of his own devising. In billing his voyage as 'a race against the clippers', he was pitting a very small boat against one of history's most successful commercial fleets. Most famous among the clippers was *Cutty Sark*. Built in 1869 for London's Jock Willis Shipping Line, she was, for decades, one of the fastest ships in the world. At 64.6m (212ft), *Cutty Sark*'s hull speed was around 18 knots and she also had an enormous sail area, a staggering 2973 sq metres (32,000 sq ft). The most Chichester could hope to wring from *Gipsy Moth* was around 8 knots. To do this, he would need to expand *Gipsy Moth*'s standard sail area, to a maximum 139 sq metres (1500 sq ft), by employing a spinnaker, a large, additional sail forward of the mainsail. Managing all this would double a workload that, in heavy seas, was certain to be tiring and dangerous.

In addition to those design flaws Chichester identified during testing, others only revealed themselves far out at sea. One was a tendency for the boat to 'broach', to suddenly change course under the influence of a strong, unexpected wind, and where the use of the rudder was insufficient to counteract the forces applied by the sails.

On the early leg of the voyage, though, these flaws didn't make themselves felt. Sailing first past southern Europe and towards the west coast of Africa, Chichester's route was positively serene. On 17 September, somewhere off the Azores, he celebrated his 65th birthday. He marked the event by donning the green velvet smoking jacket he'd packed specially for the occasion and by toasting his own health, with cocktails composed of brandy and champagne. There were presents too, including a pair of silk pyjamas from his wife, Sheila. In *Gipsy Moth Circles the World*, the bestselling account Chichester wrote of his voyage, he recalled how he had been 'intending to dine in state one night' and berated himself for forgetting his bow tie (he had to settle for a black tie instead.) 'Hammering along at a quiet 7 knots on, extraordinary pleasure, a calm, nearly flat sea... This must be one of the greatest nights of my life.' (On another occasion, his wedding anniversary, Chichester cracked open a bottle of vintage Montrachet wine.)

Below: Francis Chichester sets the course for his record-breaking single-handed circumnavigation by sail. Opposite: the flotilla that greeted *Gipsy Moth* as she sailed into Plymouth on 28 May 1967 was the largest seen here since the evacuation of Dunkirk.

For Chichester, master of understatement, such descriptions were a rare foray into loquaciousness. Not that he couldn't be funny. 'Any damn fool can navigate the world sober', he once famously said. 'It takes a really good sailor to do it drunk.' But the bulk of *Gipsy Moth Circles the World* is taken up with the necessary, if prosaic, tasks of sailing a boat single-handed; repairs to sails and how they were set, what Chichester ate, his lack of sleep and things that bumped the hull in the night.

## A NEAR CAPSIZE

After 107 days at sea, *Gipsy Moth* reached her one and only stopover, in Sydney. Chichester was, to put it mildly, glad of the break. The hull had sprung a number of leaks, while the self-steering equipment, designed to let the skipper get some well-earned rest, was also playing up. Readers of *Gipsy Moth Circles the World* often observe that in addition to the challenge of the sea, Chichester seems engaged in an ongoing battle with the boat itself. (Chichester later remarked that the ideal crew for the boat would total three: a man to navigate, an exceptionally long-armed chimpanzee to reach the sails, and an elephant to shift the impossibly stubborn tiller.) To deal with the issue of broaching, in Sydney Chichester enlisted the help of Warwick Hood, a celebrated designer of racing yachts and together they modified the boat's keel.

It would be six weeks before Chichester set sail again. Now he faced the most challenging part of the route. Within 24 hours of leaving Sydney, in the Tasman Sea, the boat suffered a partial capsize, rolling as much as 140 degrees, according to Chichester. Recounting the incident with characteristic reserve, he explained *Gipsy Moth* had been saved by her self-righting design and not for a moment did he doubt it. Such certitude proved more elusive when he encountered the Roaring Forties, the rough winds between latitudes 40° to 50° south of the equator. By the time he reached Cape Horn, the waves rose to 15m (50ft; he described them as 'great banks of grey-green earth') and the winds topped speeds of 185km/h (115mph). Upon his eventual return to England, Chichester confided to a reporter that Cape Horn was one of the few places he'd been genuinely frightened, and that a small boat only survives there with good luck.

## THE PRODIGAL RETURNS

Chichester's procession into Plymouth was almost as significant as the voyage itself. There had been no fancy website on which to monitor the boat's progress, and while the world's press had provided plenty of coverage, reports appeared days, sometimes weeks after the

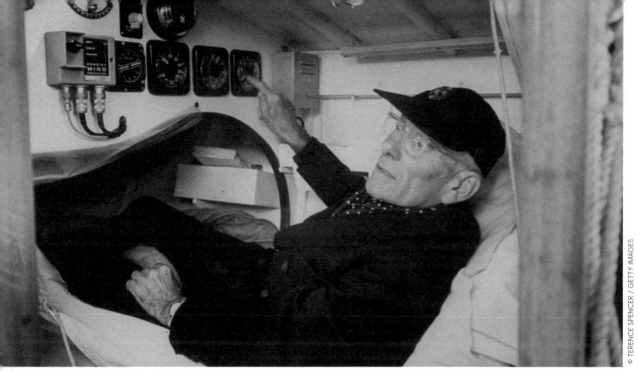

Left: Chichester in the *Gipsy Moth*'s cramped below-deck quarters. Opposite: putting *Gipsy Moth* through her paces in Plymouth Sound before embarking on the round-the-world voyage.

event. Even before Chichester's arrival, the solo circumnavigation was being hailed, with no small hyperbole, as one of the great feats of a second Elizabethan Age. In 1960, the underwater vessel, *Trieste*, had descended to Challenger Deep, at 10,916m (35,813ft) the deepest point in the ocean. Two years after Chichester's voyage, humans would land on the surface of the Moon. His achievement was only in keeping with the times.

On the evening of 28 May 1967, 119 days out of Sydney and nine months after his original departure (a total of 226 days sailing time), Francis Chichester and *Gipsy Moth IV* sailed up the Western Approaches, at a speed of a little over 4 knots. The sights that not so much greeted but gathered themselves about the record-breaking ketch were extraordinary. Hundreds of small boats surrounded *Gipsy Moth*, in what amounted to the largest flotilla to grace England's south coast since the evacuation of Dunkirk. Aircraft flew overhead, while speedboats full of Royal Marines battled through the heaving waters to provide Chichester with an official escort.

Chichester, for his part, seemed embarrassed. Kitted out not in that velvet jacket but his trusty oilskins, he pottered about on deck, busying himself with his tasks and occasionally acknowledging the crowd with a small wave. Eventually, the cluster of boats became so congested he found it difficult finding clear water. When he at last rounded the breakwater of Plymouth Sound, at 8.55pm, a signal gun from the shore announced his epic voyage was over. Having sailed 45,865km (28,500 miles), the fastest man to do so single-handedly, he was grateful for a tow from the Queen's Harbour Master. As the royal launch chuntered ahead of *Gipsy Moth*, an estimated 250,000 spectators looked on – some reports claimed twice that number – stationed on the portside or on the clifftops above, where beacons were lit. Beneath the statue of Sir

Walter Raleigh, it was surely a view to stir even the unflappable Chichester.

To the spectators who turned out in person, and to the millions watching on television, *Gipsy Moth* certainly looked like she had sailed around the world; her sails were patched and her sides streaked with rust. If Chichester appeared tired, for many he was the perfect champion; stoic rather than romantic, a conservative hero for what was still, despite the best efforts of 1960s subculture, a deeply conservative Great Britain.

Some 34 years later, in the introduction to a 2001 edition of *Gipsy Moth Circles the World*, the writer (and sailor) Jonathan Raban summed up well the mood during Chichester's televised return: 'People found something to celebrate in themselves as they watched their screens in pubs and clubs around the country', wrote Raban who, as a cool twenty-something, had been watching himself that evening. 'Their new, stooped, short-sighted knight reminded them of the salt in their veins, their brave historic past, their English mettle. Sir Francis stood as a living refutation to the seedy claims of sex, drugs, and rock and roll.'

Chichester stepped onto the dockside a little after 10pm, whereupon he was whisked away to a press reception. Asked to describe how he felt, the fellow who was taciturn at the best of times replied he had just spent the past four months in a small boat. And that coming ashore was 'like bringing a man out of a cave'.

## ARISE, SIR FRANCIS

In the decades that followed, the records Chichester set have been broken numerous times; by more advanced boats, and by those whose skippers have not required the luxury of a stopover. It is worth remembering, however, the scale of some of Chichester's achievements. For example, the 22,690 km (14,100 miles) he clocked, for the longest single-handed voyage by a small boat (Plymouth to Sydney), beat

## WORLDS APART

**Sir Francis Chichester's single-hull ketch *Gipsy Moth IV* (1966-1967)**
Waterline length: 11.7m
Sail area: 79 sq metres
Top speed: 8 knots
Duration of circumnavigation: 226 days at sea

**Dame Ellen MacArthur's trimaran *B&Q* (2005)**
Waterline length: 23m
Sail area: 373 sq metres
Top speed: 35 knots
Duration of circumnavigation: 71 days at sea

the previous record by almost double – the 11,900km (7400 miles) set by Argentina's Vito Dumas in 1943. And compared with the speed merchants who came later – the likes of Dame Ellen MacArthur who in 2005 zipped around the world in 71 days (see p.119) – Chichester spent three times longer at sea.

For *Gipsy Moth*, a waterline length of 11.7m (38ft 6in) meant reaching a speed of even 10 knots was the stuff of dreams. *B&Q*, the trimaran MacArthur used to set her record had a waterline length of 23m (75ft). Spread across three separate hulls, not only did this make *B&Q* incredibly stable, it also meant it could plane and reach speeds of 35 knots; twice as fast as *Cutty Sark*. And compared with *Gipsy Moth*'s troublesome 79 sq metres (854 sq ft) of labour-intensive canvas, MacArthur was able to call upon 373 sq metres (4013 sq ft) of high-tech, mechanically operated fabric sails. Where MacArthur relied upon state-of-the art meteorological data, Chichester fell back on a VHF radio and short-range 'radio finder', one that enabled him to calculate bearings within just a 96km (60-mile) radius. As Britain's *Daily Telegraph* noted in 2005, when reporting on MacArthur's success, Francis Chichester was 'no better equipped than Captain Cook (or Captain Pugwash)'.

Among the messages of congratulations that awaited Chichester on his return was a handwritten note from the Queen and Duke of Edinburgh. 'Welcome Home', it read and was signed, 'Elizabeth R, Philip.' In recognition of his achievements, Chichester was knighted, in July 1967, with the same sword used to bestow the honour upon Francis Drake.

Sir Francis Chichester died in 1972, five years after he became a household name and just one year after a second solo circumnavigation, this time aboard *Gipsy Moth V*. Not bad for someone once given six months to live (it was subsequently suggested that his original cancer might have been misdiagnosed). Chichester was buried near Barnstaple, in his native Devon, but honoured once more in Westminster Abbey, where his name was added to the Navigators' Memorial – alongside that of Francis Drake and James Cook.

## 'HARBINGER OF THE FUTURE'

More than 50 years after Chichester's circumnavigation of the globe, its significance abides. 'It means hardly less in the 21st century than it did in 1967', wrote Jonathan Raban, 'for Chichester (as we see only now) decisively altered the terms on which lone sailors put to sea. He has turned out not to be a sentimental throwback to the past... but a harbinger of the future.' One year later, in 1968, Britain's Robin Knox-Johnston embarked on his own round-the-world voyage, which he made solo and non-stop. Yet his journey did not fire the imagination as Chichester's had done. In part, Knox-Johnston suffered for not being first, yet also because the voyage was more leisurely. Knox-Johnston reflected on his 313 days at sea, aboard a generally agreeable ketch named *Suhaili*, as one of his life's privileges. He also had the advantage of being half Chichester's age (he was 29) and grappling with around half *Gipsy Moth*'s sail area. And while silk pyjamas and champagne cocktails are hardly the tastes of an ascetic, there was a hard-bitten drive to the curmudgeonly Chichester that had evidently appealed to the British public.

## THE RESURRECTION OF GIPSY MOTH

What of the boat Chichester so despised, and which brought him such renown? *Gipsy Moth IV* turned out to be the swansong for her designers, Illingworth and Primrose, their last ocean-going vessel. In the years following her record voyage, she was put on display, alongside *Cutty Sark*, in Greenwich in south London. The thousands who flocked to see her,

Clockwise, from left: Chichester's epic achievement captured the British public's imagination and, despite the rain, a 250,000-strong crowd turned out in Plymouth to celebrate his 1967 return; Chichester waves to the crowds after a presentation at London's Mansion House; Francis Chichester is knighted by Queen Elizabeth II.

and who walked across those slippery decks, eventually caused sufficient damage that the exhibit was closed and she fell into disrepair. *Gipsy Moth* languished until 2004, when she was resurrected and renovated in a joint project between her then custodians, the Maritime Trust, and the magazine *Yachting Monthly*. If Chichester had baulked at the original £25,000 ($32,000) price, he would have needed more than brandy and champagne when handed the bill for the refurb. Carried out at the original Camper & Nicholsons yard, the job was, in honour of Chichester's feat, carried out at cost price: £300,000 ($385,900).

Renovation of this classic boat has led to reassessment of the yacht's design. The compromises hinted at by Messrs Illingworth and Primrose, those Chichester blamed for failing to meet his brief, came partly at the behest of the man himself. '[Chichester] said it was built expressly for going solo round the world,' claimed John Walsh, the man who led the refurbishment, in an interview with the *Telegraph*. 'But it has two loos, space to sleep five and a boudoir for Lady Chichester.' It seems Sir Francis had planned for the retired boat to serve as a pleasure cruiser, perhaps to see some return on that £25,000 investment.

'Angus [Primrose] had to give him the biggest, lightest, fastest boat he could, and he did this remarkably well', reckoned Colin Silvester, custodian of the designers' archive, in 2016. Speaking to the hugely popular sailing website, Classic Boat, Silvester added: [*Gipsy Moth*] had to be light enough to keep moving in light airs, and strong enough to cope with the Southern Ocean. She had to be... driven by a sail area manageable for one man... [and] carry the waterline length necessary to generate a good hull speed.' And what of Chichester's complaint that the boat was inconsistent when heeling? 'She was meant to sail on her ear', Silvester said. 'By and large she did it.'

The project to resurrect *Gipsy Moth IV* finally saw her put to sea once more, in 2005; she sails to this day. Her successor, *Gipsy Moth V*, was not so fortunate. In December 1982, eleven years after Chichester's second round-the-world voyage, *Gipsy Moth V* was competing in a race, from Cape Town to Sydney. Her skipper, Desmond Hampton, was 1450km (900 miles) south of Melbourne when he decided to entrust his course to the boat's self-steering system, one not dissimilar to that which had given Chichester trouble aboard her predecessor. Shortly after Hampton retired to bed, *Gipsy Moth V* struck rocks, on Gabo Island, and broke up.

Hampton survived, and with the help of a lighthouse keeper salvaged much of the equipment on board the boat.

# NAVIGATION 101

How did Chichester know where he was? In a modern yacht, a tablet-sized chart plotter loaded with the latest charts for the area being navigated, combined with GPS (global positioning system) data from satellites, computes your position, course and speed with high accuracy.

However, GPS only became widely available in the 1980s. Chichester, once out of sight of land, would have used the sun and stars to find his position in much the same way as sailors over the last millennia. Whereas early navigators used sticks, discs and bits of string to measure angles between stars, the sun and the horizon, Chichester, like Captain Cook, used a sextant (below) to measure the angles to derive his position with an accuracy of up to a mile or so.

So what happens when it's cloudy and these celestial bodies are not visible? The answer is that he would have used the formula *distance = speed x time*, combined with the boat's compass course, to calculate how far the boat had sailed, and in what direction, from the last known position. This technique is called 'dead reckoning'. To establish their speed, early sailors would throw a chip of wood into the water at the bow and time how long it took to pass level with the stern. Chichester used a hull-mounted paddle-wheel speed log, backed up by a torpedo-shaped impeller-type log, which is towed behind the boat.

This sounds straightforward – but try doing navigation when you're freezing cold, the wind is over 50 knots and waves higher than the mast are tossing the boat around like a cork for days on end, and you can see why Chichester was knighted.

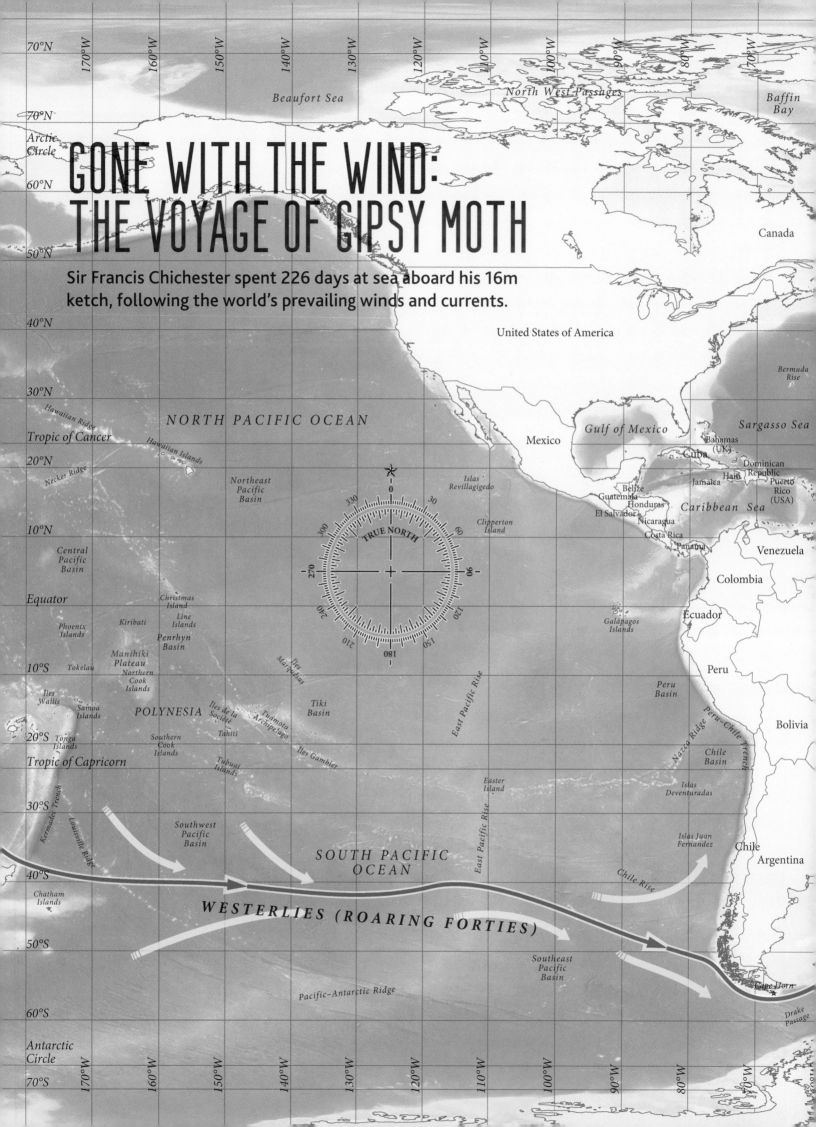

# GONE WITH THE WIND: THE VOYAGE OF GIPSY MOTH

Sir Francis Chichester spent 226 days at sea aboard his 16m ketch, following the world's prevailing winds and currents.

# MAYDAY! THE PERILS OF THE HIGH SEAS

## CARGO SHIPS

Huge, unwieldy vessels with small crews operating minimal watches may be unlikely to spot a small yacht.

## BOOM STRIKE

Among the most common sailing injuries. Freak weather or waves, or poorly rigged sails can cause the boom to swing violently.

## MAN O

Rescuing
high seas
so, when
sailors st
their yach

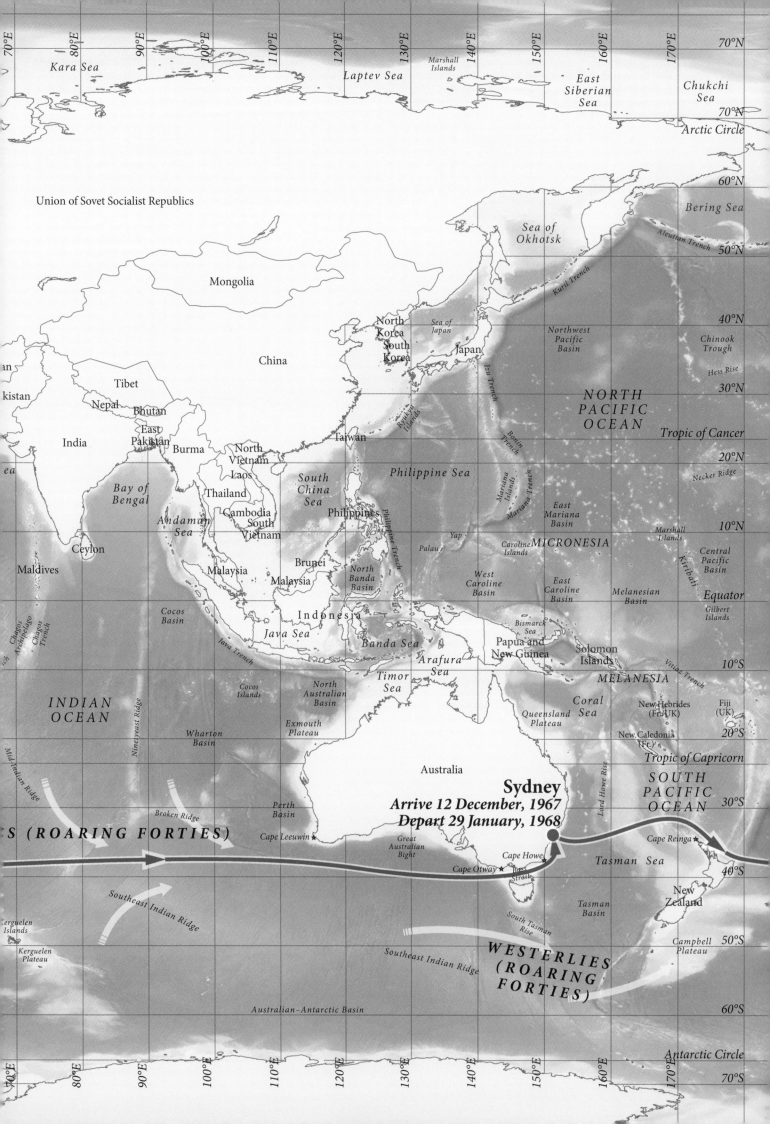

## WAVES

The height of a wave need only be a third of a boat's length to roll it over; or two thirds to overwhelm it entirely.

## PIRATES

Over the last decades, the western Indian Ocean, the Caribbean and SE Asian waters have all seen pirate attacks.

## ...ERBOARD

...'MOB' on the ... very difficult; ...n deck, many ...y clipped to ... via life-lines.

## WHALES

As racing yachts get faster, whale strikes are said to be increasing – keels and rudders are especially vulnerable.

## SEA CONTAINERS

With ever-busier shipping lanes, the threat to a yacht from a semi-submerged container, and other large flotsam, is considerable.

# 'I'D SEEN AN ELECTRIC STORM OFF MADAGASCAR, BUT NOT LIKE THIS'

**Joshua Slocum was the first man to sail solo around the world. In an extract from his 1899 account of the voyage, he describes the perilous final days of his journey.**

On June 5, 1898, the *Spray* sailed for a home port, heading first direct for Cape Hatteras [off North Carolina]. On the 8th of June she passed under the sun from south to north; the sun's declination on that day was 22 degrees 54', and the latitude of the *Spray* was the same just before noon. Many think it is excessively hot right under the sun. It is not necessarily so. As a matter of fact the thermometer stands at a bearable point whenever there is a breeze and a ripple on the sea, even exactly under the sun. It is often hotter in cities and on sandy shores in higher latitudes.

The *Spray* was booming joyously along for home now, making her usual good time, when of a sudden she struck the horse latitudes, and her sail flapped limp in a calm. I had almost forgotten this calm belt, or had come to regard it as a myth. I now found it real, however, and difficult to cross. This was as it should have been, for, after all of the dangers of the sea, the dust-storm on the coast of Africa, the 'rain of blood' in Australia, and the war risk when nearing home, a natural experience would have been missing had the calm of the horse latitudes been left out. Anyhow, a philosophical turn of thought now was not amiss, else one's patience would have given out almost at the harbor entrance. The term of her probation was eight days. Evening after evening during this time I read by the light of a candle on deck. There was no wind at all, and the sea became smooth and monotonous. For three days I saw a full-rigged ship on the horizon, also becalmed.

Sargasso, scattered over the sea in bunches, or trailed curiously along down the wind in narrow lanes, now gathered together in great fields, strange sea-animals, little and big, swimming in and out, the most curious among them being a tiny seahorse which I captured and brought home preserved in a bottle. But on the 18th of June a gale began to blow from the southwest, and the sargasso was dispersed again in windrows and lanes.

On this day there was soon wind enough and to spare. The same might have been said of the sea. The *Spray* was in the midst of the turbulent Gulf Stream itself. She was jumping like a porpoise over the uneasy waves. As if to make up for lost time, she seemed to touch only the high places. Under a sudden shock and strain her rigging began to give out. First the main-sheet strap was carried away, and then the peak halyard-block broke from the gaff. It was time to reef and refit, and so when 'all hands' came on deck I went about doing that.

The 19th of June was fine, but on the morning of the 20th another gale was blowing, accompanied by cross-seas that tumbled about and shook things up with great confusion. Just as I was thinking about taking in sail the jibstay broke at the masthead, and fell, jib and all, into the sea. It gave me the strangest sensation to see the bellying sail fall, and where it had been suddenly to see only space. However, I was at the bows, with presence of mind to gather it in on the first wave that rolled up, before it was torn or trailed under the sloop's bottom. I found by the amount of work done in three minutes' or less time that I had by no means grown stiff-jointed on the voyage; anyhow, scurvy had not set in, and being now within a few degrees of home, I might complete the voyage, I thought, without the aid of a doctor. Yes, my health was still good, and I could skip about the decks in a lively manner, but could I climb? The great King Neptune tested me severely at this time, for the stay being gone, the mast itself switched about like a reed, and was not easy to climb; but a gun-tackle purchase was got up, and the stay set taut from the masthead, for I had spare blocks and rope on board with which to rig it, and the jib, with a reef in it, was soon pulling again like a 'sodger' for home. Had the *Spray's* mast not been well stepped, however, it would have been 'John Walker' when the stay broke. Good work in the building of my vessel stood me always in good stead.

On the 23rd of June I was at last tired, tired, tired of baffling squalls and fretful cobble-seas. I had not seen a vessel for days and days, where I had expected the company of at least a schooner now and then. As to the whistling of the wind through the rigging, and the slopping of the sea against the sloop's sides, that was well enough in its way, and we could not have got on without it, the *Spray* and I; but there was so much of it now, and it lasted so long! At noon of that day a winterish storm was upon us from the nor'west. In the Gulf Stream, thus late in June, hailstones were pelting the *Spray*, and lightning was pouring down from the clouds, not in flashes alone, but in almost continuous streams. By slants, however, day and night I worked the sloop in toward the coast, where, on the 25th of June, off Fire Island, she fell into the tornado which, an hour earlier, had swept over New York city with lightning that wrecked buildings and sent trees flying about in splinters; even ships at docks had parted their moorings and smashed into other ships, doing great damage. It was the climax storm of the voyage, but I saw the unmistakable character of it in time to have all snug aboard and receive it under bare poles. Even so, the sloop shivered when it struck her, and she heeled over unwillingly on her beam ends; but rounding to, with a sea-anchor ahead, she righted and faced out the storm. In the midst of the gale I could do no more than look on, for what is a man in a

Left: contemporary drawing of the *Spray* moored at Gibraltar.

storm like this? I had seen one electric storm on the voyage, off the coast of Madagascar, but it was unlike this one. Here the lightning kept on longer, and thunderbolts fell in the sea all about. Up to this time I was bound for New York; but when all was over I rose, made sail, and hove the sloop round from starboard to port tack, to make for a quiet harbor to think the matter over; and so, under short sail, she reached in for the coast of Long Island, while I sat thinking and watching the lights of coasting-vessels which now began to appear in sight. Reflections of the voyage so nearly finished stole in upon me now; many tunes I had hummed again and again came back once more. I found myself repeating fragments of a hymn often sung by a dear Christian woman of Fairhaven when I was rebuilding the *Spray*. I was to hear once more and only once, in profound solemnity, the metaphorical hymn:

> By waves and wind I'm tossed and driven.

And again:

> But still my little ship outbraves
> The blust'ring winds and stormy waves.
> After this storm I saw the pilot of
> the *Pinta* no more.

The experiences of the voyage of the *Spray*, reaching over three years, had been to me like reading a book, and one that was more and more interesting as I turned the pages, till I had come now to the last page of all, and the one more interesting than any of the rest.

When daylight came I saw that the sea had changed color from dark green to light. I threw the lead and got soundings in thirteen fathoms. I made the land soon after, some miles east of Fire Island, and sailing thence before a pleasant breeze along the coast, made for Newport. The weather after the furious gale was remarkably fine. The *Spray* rounded Montauk Point early in the afternoon; Point Judith was abeam at dark; she fetched in at Beavertail next. Sailing on, she had one more danger to pass – Newport harbor was mined. The *Spray* hugged the rocks along where neither friend nor foe could come if drawing much water, and where she would not disturb the guard-ship in the channel. It was close work, but it was safe enough so long as she hugged the rocks close, and not the mines. Flitting by a low point abreast of the guard-ship, the dear old *Dexter*, which I knew well, some one on board of her sang out, 'There goes a craft!' I threw up a light at once and heard the hail, '*Spray*, ahoy!' It was the voice of a friend, and I knew that a friend would not fire on the *Spray*. I eased off the main-sheet now, and the *Spray* swung off for the beacon-lights of the inner harbor. At last she reached port in safety, and there at 1am on June 27, 1898, cast anchor, after the cruise of more than forty-six thousand miles round the world, during an absence of three years and two months, with two days over for coming up.

**Left:** illustration of Slocum conversing on the *Spray's* deck from his book, *Sailing Alone Around the World*.

KRYSTYNA
CHOJNOWSKA-
LISKIEWICZ

## 1976-78

### KRYSTYNA CHOJNOWSKA-LISKIEWICZ
#### FIRST WOMAN TO SAIL AROUND THE WORLD

A name not widely known beyond yachting circles, Poland's Krystyna Chojnowska-Liskiewicz entered the sailing limelight as the winner of a competition. In 1975, the Polish Sailing Association decided to celebrate that year's United Nations' International Women's Year by sponsoring a voyage around the world. Then aged 40, Chojnowska-Liskiewicz had learned to sail on the lakes of Ostróda, a well-known Polish yachting destination. In 1976, she sought to repeat a route first taken by Joshua Slocum between 1895 and 1898. Aboard her yacht, *Mazurek*, Chojnowska-Liskiewicz set sail on 28 March from the Canary Islands.

She returned, some 401 days later, on 21 April 1978, having logged 50,156km (31,166 miles). Chojnowska-Liskiewicz went on to become a marine engineer and was inducted into New York City's Explorers Club, the US equivalent of the Royal Geographical Society.

From Joshua Slocum to Sir Francis Chichester, you might be forgiven for thinking the history of sailing around the world is an all-male affair. But for the best part of 50 years, women have been staking all on the high seas. Here, we salute the pioneers.

# MAKING

1977

1977-78

DAME NAOMI JAMES

### CLARE FRANCIS
#### FIRST FEMALE SKIPPER, WHITBREAD ROUND THE WORLD

Despite training as a ballerina, by the age of 27 the Surrey-born Francis found herself in a dead-end office job. Deciding life needed more excitement, in 1973 she sailed single-handedly across the Atlantic in a 9.8m (32ft) boat; the voyage lasted 37 days. Later that year, she entered the Whitbread Round Britain Race, finishing third. With a reputation as a fierce competitor, in 1977 she became the first woman to skipper a yacht in the Whitbread. At the helm of *ADC Accutrac*, her crew finished fifth, in a race won by legendary Dutch sailor, Cornelis 'Conny' van Rietschoten, the only captain to win the Whitbread twice.

On her retirement from sailing, Francis decided to try and make a living as a writer. Her determination proved the springboard for a successful career, and she went on to become a bestselling author. To date, Francis has written 12 novels, but her first publications were non-fiction accounts of her aquatic adventures: *Come Hell or High Water* (1977); *Come Wind or Weather* (1978); and *The Commanding Sea* (1981).

### DAME NAOMI JAMES
#### FIRST SOLO CIRCUMNAVIGATION VIA CAPE HORN

Born in New Zealand, Naomi James was raised on a sheep farm and trained as a hairdresser. During a visit to Europe in 1975, she met Rob James, her future husband and a skipper for renowned British sailor Chay Blyth. After basic sailing lessons, James announced her ambition to sail around the world. Not only did she seem ill-prepared but, possibly, recklessly ambitious. The route she chose was eastwards, that of Sir Francis Chichester in 1966; and, crucially, round Cape Horn and through the perilous Southern Ocean. On 9 September 1977, James set sail from Dartmouth, Devon, aboard the yacht *Express Crusader*, a loan from Blyth. After 272 days at sea, she returned to Dartmouth on 8 June 1978 – starting and finishing in the English Channel was then a requirement for speed records.

James was the first woman to circumnavigate via Cape Horn and was narrowly beaten to the overall record (by Poland's Krystyna Chojnowska-Liskiewicz). Made a Dame in 1979, James eventually gave up sailing due to seasickness.

# WAVES

## 1987-88

## 2005-06

### KAY COTTEE
#### FIRST SOLO NON-STOP CIRCUMNAVIGATION

Unlike Naomi James, Australia's Kay Cottee hailed from a long line of sailors, having grown up on the shores of Sydney's Botany Bay. Nonetheless, when she set out from Sydney Harbour on 29 November 1987, Cottee had set herself a formidable challenge; to be the first woman to sail single-handedly and non-stop around the world. Compared with the corporate-funded voyages that came later, hers was a small-scale affair – to buy her boat, *Blackmores First Lady*, she relied on donations from family and friends and also resorted to selling many of her possessions. In the coming years, Cottee's route was adopted by numerous sailors in pursuit of records, although it undoubtedly carries risks. In the Southern Ocean, she was repeatedly knocked off her feet, was washed overboard and also capsized; luckily the boat rolled through 360 degrees. In calmer seas, she relaxed by knitting and baking. Unaware of the extensive news coverage during her 189 days at sea, Cottee returned to a crowd of 10,000 well-wishers and was subsequently awarded the Order of Australia.

### DEE CAFFARI, MBE
#### FIRST WOMAN TO SAIL SOLO AND WESTWARD AROUND THE WORLD

It was Chay Blyth who hit upon the idea of sailing the 'wrong way' around the world: westward, against the prevailing winds and currents. He was also the first to do so, in 1970–71. Dee Caffari was an accomplished sailor when she set out to become the first woman to complete the same voyage single-handed, but her plan was ambitious, given she had done little solo sailing. Caffari set out from Portsmouth on 20 November 2005. Ahead lay a journey of 46,800km (29,100 miles), with none of the respite that following winds offer those sailing in the standard direction. Instead, during her 178 days at sea, she endured 34 days of gale-force weather, as well as more familiar problems: icebergs, doldrums and tiredness. Her autopilot broke down on Christmas Day and Caffari later confessed there were several occasions when she thought she wouldn't finish (or even survive). But finish she did, on 18 May 2006. Three years later, in February 2009, Caffari took part in the Vendée Globe, the four-yearly, solo race around the world. On its completion, she set yet another record, as the first woman to circumnavigate by boat, in both directions.

KAY
COTTEE

DEE
CAFFARI

# BY BOAT

**1989**

**2005**

DAME ELLEN
MACARTHUR

### TRACY EDWARDS
### FIRST ALL-FEMALE CREW TO COMPLETE WHITBREAD

Few sailors have proved as single-minded as Tracy Edwards. She entered her first Whitbread Round the World Race in 1985, aged 23, as the boat's cook. By 1989, aboard the *Maiden*, she was skippering the Whitbread's first all-female crew. At a time when the sport was dominated by men, her kinder critics suggested Edwards was unqualified. The less kind doubted she'd finish the first leg, or claimed she was endangering the lives of her 12-strong crew. But the *Maiden* defied the sceptics, notching impressive wins in two legs, the best result for a British boat for a decade. Edwards spent the next 10 years trying to better her Whitbread success. In 1998, she entered the Jules Verne Trophy, a race for the fastest circumnavigation. The all-female crew, sailing the *Royal & SunAlliance*, were on course to break the record when heavy weather off Chile broke the boat's mast; they took 16 days to reach land. Edwards' subsequent fortunes ebbed and flowed: in 2005, she was declared bankrupt; in 2018, she featured in *Maiden*, an award-winning film about the 1989 race. She also restored the boat and reunited the crew.

TRACY
EDWARDS

### DAME ELLEN MACARTHUR
### FASTEST SOLO CIRCUMNAVIGATION

Ellen MacArthur was born in landlocked Derbyshire but, by 2005, had relocated to Cowes on the Isle of Wight, home to one of the world's most famous sailing regattas. On 28 November 2005, she set sail from Falmouth, at the helm of *B&Q*, a state of-the-art, 23m (75ft) trimaran. In her sights was the record for the fastest solo circumnavigation, of 72 days, 22 hours, 54 minutes and 22 seconds, set by Francis Joyon in 2004. The boat, designed specifically for MacArthur, was 4.5m (15ft) shorter than Joyon's yacht, *IDEC*, enabling it to surf waves for longer periods. *B&Q*'s electronics also included the latest satellite navigation systems, continually fed by meteorological data from two separate weather stations, based in the US and Germany. Both systems were linked to the boat's autopilot, effectively allowing *B&Q* to correct its own course (but only when MacArthur was required elsewhere, such as mending sails or, heaven forbid, sleeping.)

Film footage captured during the voyage depicts an exhausting experience. Among the many obstacles MacArthur had to overcome were icebergs and extreme winds in the Southern Ocean, a lack of wind across the Atlantic and technical problems that seemed to occur almost daily. On one occasion, she was forced to scale the boat's 27m (90ft) mast to fix a problem with the main sail; another time, sailing at speed, she came close to hitting a surfacing whale.

MacArthur crossed the finishing line, off Ushant, in France, to set a new record. In completing 40,230km (25,000 miles) in 71 days, 14 hours, 18 minutes and 33 seconds, she lowered Joyon's mark by one day, eight hours, 35 minutes and 49 seconds. Her average speed, of 15.9 knots, was almost double that of Sir Francis Chichester's *Gipsy Moth IV*, in 1966. In addition to the circumnavigation record, MacArthur's voyage also poached another from Joyon; the fastest time to the equator, around the Capes (of Good Hope, Leeuwin and Horn) and back to the equator once more.

# ALL AT SEA: THE TRAGIC TALE OF DONALD CROWHURST

**It's one of the most haunting stories of modern sailing: how a weekend amateur found himself 'leading' the first solo, round-the-world race – and then disappeared.**

In 1967, Francis Chichester was knighted for circumnavigating the world by the clipper route. Such was public interest in his success that, the following March, Britain's *Sunday Times* newspaper sought to capitalise on Chichester's success by sponsoring an around-the-world event of its own. The inaugural Golden Globe Race offered two prizes: a trophy, and attendant prestige, for the winner; and a separate cash pot, of £5000 ($6400), for the fastest circumnavigation, be it the first boat home or the last.

The rules were loose, stipulating the voyage could start at any time between 1 June and 31 October 1968, but had to be made single-handed and also non-stop. Given Chichester had made port, once, in Sydney, this was a feat yet to be accomplished.

And that was it. There was no entry fee and competitors were free to set sail as soon as they were ready. Chichester could not be persuaded to participate – although he served as a race judge – but other celebrated yachtsmen were only too willing to accept, not least because a record-setting non-stop voyage was already in their sights. Those who put to sea included Chay Blyth, John Ridgway and Robin Knox-Johnston.

## THE TEIGNMOUTH ELECTRON SETS SAIL

On 31 October, the final day of entry, the last competitor in the race set sail, aboard a purpose-built trimaran, the *Teignmouth Electron*. Its skipper Donald Crowhurst was not a name familiar to the yachting world; in truth, it was a name barely known to sailors of any kind.

Crowhurst's principal connection with the sport was tenuous to say the least: his business manufactured navigational equipment. By 1968, however, it was in serious financial trouble. Born in British India in 1932, and raised on Kipling, Crowhurst had been stirred by Chichester's achievement, not to mention the cash prize. In his now home of Teignmouth, he persuaded an investor, Stanley Best, to fund a boat to take part in the Golden Globe Race. If his attempt at a circumnavigation was unsuccessful, Crowhurst promised Best he would buy the boat himself. As an additional show of faith, Crowhurst offered up his family home as collateral.

# BY BOAT

To keep the organising committee and press updated on their progress, competitors in the Golden Globe were required, at various locations, to radio in their positions. Crowhurst's initial reports were mixed but, on occasion, very promising; in the early stages of the race, for example, the unheralded sailor reported covering 450km (280 miles) in a single day, a distance that smashed the then world record. Throughout November and December, things began to change. Crowhurst's reports became vague – eventually they stopped altogether.

## REALITY BITES

The press started to speculate as to Crowhurst's fate. Was he still in the race? Was he even alive? The answers were 'no' and 'yes'. In truth, Crowhurst had begun to experience problems almost immediately. *Teignmouth Electron* leaked badly and, compared with the Bristol Channel, where he occasionally sailed at the weekend, navigation on the open ocean was a struggle. Somewhere in the Atlantic, it dawned on Crowhurst that a circumnavigation of the world was impossible; given the state of his boat, he noted in his log that he doubted he would even survive the Southern Ocean. Turning back, however, was not an option; he would have to buy the boat from Best, with money he didn't have, and he would also lose his house.

So, he concocted a plan. Rather than sail east, to the Southern Ocean and probable death, Crowhurst made instead for South America, where it is believed he put in somewhere along the Brazilian coast. He made much-needed repairs, illegal under Golden Globe rules. It was sometime in early December 1968 that Crowhurst began reporting false positions; claiming to be off the Cape of Good Hope, bound for Australia. He compounded his deception with equally fictitious logbooks.

Crowhurst's plan, it seems, was to potter about the southern hemisphere. Once the race had rounded Cape Horn, Crowhurst would rejoin it, trailing home at the back of the field, a loser of

the race but a success in the eyes of the wider public. Being ineligible for any prize, Crowhurst reckoned the race judges were unlikely to check his logs very closely.

On 22 April 1969, Robin Knox-Johnston sailed back into Falmouth and won the race. He entered the history books as the first man to sail around the world solo and non-stop. All Crowhurst needed to do now was to time his run, but events conspired against him. Crowhurst had set sail the previous October, Knox-Johnston nearly four months before him; it was assumed therefore that Crowhurst would easily beat Knox-Johnston's time, of 312 days. Only two sailors were ahead of Crowhurst, Bernard Moitessier and Nigel Tetley. The former, a free-spirited Frenchman, had done the hard part, battling through the Southern Ocean, only to protest at what he saw as the gross commercialisation of sailing by abandoning the race and heading for Tahiti. Tetley, in contention for the fastest time but believing Crowhurst still posed a genuine challenge, drove his own boat so hard it fell apart, necessitating a Mayday call and no option but to abandon.

## THE FINAL LOG ENTRY

The £5000 prize seemingly belonged to Crowhurst. Except for one, unavoidable fact: if he returned home, the star of a story that had occupied newspaper front pages for months, the race judges were guaranteed to check his logbooks carefully; and among the judges would be Sir Francis Chichester, a known stickler. The last entry in Crowhurst's log was made on 1 July 1969. Earlier entries revealed evidence of his falsified positions, along with musings that suggested a man under significant pressure. 'It is finished', Crowhurst wrote on the last page of his logbook. 'It is finished, it is the mercy.' His boat was found, abandoned, nine days later, although his body was never recovered. The £5000 for the fastest circumnavigation was duly awarded to Robin Knox-Johnston. He, in turn, gave it to Crowhurst's family.

# HOW TO RUN AWAY TO SEA — AND MAKE IT COUNT

**It wasn't the oceans or even her parents that nearly sank Laura Dekker's attempts to become the youngest marine circumnavigator – it was the Dutch childcare authorities.**

Laura Dekker's family dynamic probably explains a lot. In May 2009, the teenager from Utrecht set sail on her first solo voyage, from the Netherlands, aged just 13. She made it across the English Channel before deciding to quit. The coastguard in the port of Lowestoft picked her up but her father, Dick, initially refused to collect her. His daughter had made the outward voyage alone, he told the British authorities, so she was perfectly capable of returning alone.

It was then the Dutch courts intervened. Childcare agencies in the Netherlands brought a legal action, Dekker was made a ward of court and, despite support from her parents, banned from sailing. The following year, aged 14 but with the ban still in effect, Dekker went anyway. Having withdrawn €3500 ($3890) from her bank account, Dekker set off aboard her yacht, *Guppy*, on 17 December. She made it as far as Sint Maarten, in the Dutch Antilles, before a Europe-wide search – given Dekker's form it was hardly a panic – caught up with her four days later.

As the press noted at the time, Laura Dekker was born to sail. Born on a yacht, in fact, moored in the port of Whangarei, in New Zealand, where her parents had put in during their own seven-year voyage around the world. Dekker enjoyed NZ citizenship as a consequence and, as the legal action dragged on, she entertained plans to sail from New Zealand – overtures made in that direction, however, received little encouragement.

Respite came when, in July 2010 and following a medical assessment, the ban on Dekker was lifted; Dutch child protection agencies were in the process of getting the original ban extended only

for the courts to rule against them. Free to sail, once she kept up with her schoolwork, Dekker finally set out, from Portugal, on 21 August 2010.

## FREE AT LAST

Her westward voyage, of 518 days and 10,260km (6380 miles), included one stretch of 47 days, alone and non-stop, from Darwin, Australia to the South African port of Durban. Dekker claimed subsequently that she enjoyed the solitude of single-handed sailing, and in particular her encounters with penguins and dolphins. The weather was mainly favourable, though her voyage was not without incident. There was a near-miss with a cargo ship in the Panama Canal, concerns about marauding pirates in the Indian Ocean, and flying fish that landed on her deck and required rescue. Conditions proved poor around the Cape of Good Hope, where the waves reached 5.5m (18ft) and *Guppy*'s sails were badly damaged.

On the date of her return, 21 January 2012, Dekker was 16 years and 123 days old. The record she claimed, for the youngest person to sail around the world, had been hotly contested. The previous holder, Australia's Jessica Watson, had been three days shy of her 17th birthday when she set the record in 2010; American sailor Abby Sunderland was 16 when she attempted to beat Watson's record that same year. The record had previously been held by Sunderland's brother, Zac, who set the mark, subsequently beaten by a 17-year-old Briton, Mike Perham, in August 2009.

Authorities of a different kind continued to check Dekker's ambition. In a bid to discourage yachting youths in future, Guinness World Records do not recognise circumnavigations made by minors.

# A RACE FOR MAD MEN: THE VENDÉE GLOBE

**It already had a forbidding reputation: a single-handed, non-stop race around the planet. And then, in the 1996-97 edition, colossal storms struck the Southern Ocean and the world watched as rescuers struggled to save sailors fighting for their lives in huge seas.**

No yacht race can match the single-minded, single-handed lunacy of the Vendée Globe. Held every four years, it requires a special kind of sailor to endure the voyage, solo and non-stop around the world, through the treacherous waters of the Southern Ocean where, as a famous edition of the race demonstrates, competitors are never more than one wave away from catastrophe.

## RUNNING OUT OF AIR

January 1997: Floating in the freezing waters of the Southern Ocean, the air in his upturned yacht fast disappearing, Tony Bullimore doubtless wished he'd stuck to reggae. Originally from Southend, Essex, in the 1960s Bullimore had opened the Bamboo Club, in Bristol; along with his Jamaican wife, Lalel, the couple had welcomed artists such as Bob Marley and Tina Turner.

After music, Bullimore's other great passion was sailing. A proficient yachtsman – despite subsequent efforts by the British press to portray him as hapless – Bullimore had been participating in single-handed races since the early 1970s. In 1985, he won the prestigious Round Britain Race and was named yachtsman of the year. The same year, in fact, that Simon Le Bon, of Duran Duran, was rescued from his own boat, *Drum*, after it capsized during the Fastnet Race.

Compared with Fastnet, a 965km (600-mile) out-and-back dash from Cowes, on the south coast of England, to Ireland, the Vendée Globe is a very different beast. Non-stop and single-handed, around the world, through the most difficult oceans. Bullimore had begun the race the previous November, setting sail from the French port of Les Sables d'Olonne, aboard his 18m (60ft) racing yacht, *Exide Challenger*.

Even by the Vendée's punishing standards, the 1996–97 edition was brutal. So bad were conditions that two of the 16-strong fleet

Left: the frigid Southern Ocean is known for its rough waters, but the punishing conditions here during the 1996–97 Vendée Globe led to three capsized yachts and the dramatic rescue of their lone captains.

got no further than the Bay of Biscay. From there, the race skirted Africa, shot across the Indian Ocean to Australia, then on to the most challenging part of the course, the monstrous Southern Ocean. The event leaders were already on their way to the relative calm of the Atlantic when, at Christmas, an unusually heavy storm hit the Southern Ocean, catching out the remaining competitors. With Antarctica less than 1600km (1000 miles) to the south, the Vendée's participants caught in the storm were lashed by freezing winds and hampered by monster waves of up to 9m (30ft) high.

## HANGING ON FOR DEAR LIFE

One of the first to get into trouble was Raphael Dinelli, skipper of the *Algimouss*. His boat capsized on Christmas Day and he sent out a Mayday signal. He was rescued, two days later, by Britain's Pete Goss, aboard *Aqua Quorum*. Goss' race to reach his fellow competitor played out on news broadcasts in both Britain and France and he arrived in the nick of time, to find Dinelli standing on the hull of his rapidly sinking boat. (Goss only found the Frenchman thanks to a spotter plane from the Royal Australian Air Force.)

But the drama was only just beginning. Another sailor, France's Thierry Dubois, skipper of the yacht *Amnesty International*, had capsized around the same time as Dinelli. By now, an Australian frigate, HMAS *Adelaide*, was combing the area, a stretch of water 2575km (1600 miles) off the Australian coast. Having sent out his own distress signal, Dubois was also located, but only after numerous passes by the Aussie spotters; unable to haul himself aboard his stricken vessel, Dubois was clinging to it for dear life, and it took several attempts before it was possible to haul him safely aboard the deck of the *Adelaide*.

If it seemed the Southern Ocean was gradually shortening the odds, Bullimore's experience appeared to confirm it. *Exide Challenger* was the third vessel to activate her distress signal, and she was the last to be found. Bullimore and Dubois had been just 97km (60 miles) apart when the storm hit, but it took a further four days for the *Adelaide* to find the *Exide Challenger*. Her keel was badly damaged and, most worryingly, the Englishman was nowhere to be seen. Unable to send out a launch in the dreadful conditions, the crew of the *Adelaide* knew there were two possibilities; Bullimore was either dead or, by some miracle, trapped beneath the boat. A phone call to the yacht's designers confirmed *Exide's* internal dimensions; if luck was on his side, Bullimore might have enough air to survive for around five days.

## THE LAST-MINUTE RESCUE

With time running out, the *Adelaide* at last sent an inflatable, manned with rescue divers, alongside the upturned boat. To their relief, when divers pounded on the hull, a hand inside pounded back. Bullimore was, by now, both weak and injured. Having eaten no more than an emergency chocolate bar, and with precious little fresh water, he was suffering from hypothermia and trench foot. He was also bleeding heavily, having lost a finger when cutting rope. The naval divers could only do so much. If Bullimore wanted out, he would have to swim.

Somehow, Bullimore summoned the strength. Plucked from the sea by Chief Petty Officer Peter Wicker of the Australian navy, he joined Dubois aboard the *Adelaide* – the first thing he asked for was a cup of tea. Back in Fremantle, Western Australia, the two sailors were greeted by a large crowd and a navy band – playing Louis Armstrong's *It's a Wonderful World*. If his body was not quite intact, his nightclub manager's instincts certainly were. Following his reception on the Fremantle quayside, Bullimore reputedly sold exclusive rights to the reunion with his wife to an Australian TV network. Conducted at the British consul in Perth, the emotional event commanded a fee rumoured to be £75,000 ($97,000).

The 1996–97 edition of the Vendée Globe passed into race folklore. It was eventually won, by France's Christophe Augin aboard *Geodis*, who set a then race record of 105 days, 20 hours and 31 minutes. Of the 16 starters, just six finished; of the 10 non-finishers, Bullimore was one of six who capsized. A seventh sailor, the Canadian Gerry Roufs had, like Bullimore, got into difficulties; unlike the Englishman, contact was lost and Roufs disappeared. His last known position was south of Easter Island. The wreckage of Roufs' boat, *Groupe LG2*, was located the following June, in waters off the coast of Chile. Roufs was declared lost at sea.

## UNDER SAIL ONCE MORE

In a heartening postscript, Raphael Dinelli and his saviour Pete Goss remain friends to this day. (In August 1997, Goss was best man at Dinelli's wedding.) And despite his traumatic Vendée Globe experience, Tony Bullimore not only sailed again but competed in subsequent ocean races and record attempts. In 2007, aboard the yacht *Blue Ocean Wireless*, he attempted to beat the 2005 solo circumnavigation record set by Ellen MacArthur. Bullimore failed to conquer MacArthur's record but his voyage was best remembered for the 11 days he spent out of radio contact, leading the British press to once again question his credentials and portray him, unfairly, as a serial blunderer.

Clockwise, from top: choppy waters at the start of the 2016 Vendée Globe; the *Exide Challenger* on 7 January 1997, two days after it capsized with its skipper trapped under the hull; Bullimore at the helm of the *Team Legato*, which he captained for The Race circumnavigation challenge in 2001; Walter Greene and Tony Bullimore aboard the *Apricot* during the 1986 TwoSTAR transatlantic race.

# ALL ABOARD WITH GRETA

Since 2013, Elayna Carausu and Riley Whitelum (and baby Lenny) have been crowdfunding and vlogging their yachting adventures – including a voyage with Greta Thunberg.

In the wake of Ferdinand Magellan, Sir Francis Chichester and Ellen MacArthur, the age of social media has, naturally, produced its own generation of maritime adventurers.

Foremost among them are Riley Whitelum and Elayna Carausu. The two Australians – Whitelum is from Adelaide and Carausu from Geraldton – met in 2013, following respective careers as an oil-rig worker and a singer. Following a serious bodyboarding accident, Whitelum decided on a change of lifestyle. He travelled to Europe, purchased a boat, *La Vagabonde*, and with

only 10 hours of sailing under his belt, set sail for Italy, where he met Carausu.

The couple have been crowdfunding a voyage around the world ever since – they're in no particular hurry. In contrast to Magellan, who required the deep pockets of the king of Spain to sub his voyages, the two enterprising Aussies are using the funding site, Patreon. There they invite 'crew members' to join them (virtually, of course) by offering a range of subscriptions (they currently have nearly 3500 Patreon funders). Posting regular video content on their dedicated YouTube channel (by late 2019 they had 1.15 million YouTube subscribers), Whitelum and Carausu have established a sufficiently lucrative social media career to keep

# BY BOAT

Clockwise from top left: Riley Whitelum and Elayna Carausu aboard *La Vagabonde*; the yachting life has its upsides; portside with Greta Thunberg; vlogging *en famille*.

their travel plans afloat. They have made two Atlantic crossings and one Pacific.

In November 2019 they responded to activist Greta Thunberg's call for help to get her from the US to a UN climate summit in Madrid – welcoming such a high-profile passenger on board *La Vagabonde* is bound to grow their audience even more.

The images and short films they upload across various platforms range from breathtaking views to the tedium of everyday life at sea – even beautiful yachtspeople need to do laundry and boat maintenance. After seven years and counting, they've had a baby, Lenny, and diversified their content, adding parenting advice to an online library of sailing and fishing guides.

# FOOT BY

The first verified walk around the world was completed, in some ways, as an act of remembrance. Starting out from Minnesota, in 1970, with his brother, John, the USA's Dave Kunst spent the next four years traipsing a total of 23,255km (14,450 miles) across four continents. Taking with them a mule (named 'Willie Makeit'), the Kunsts walked first across the US before, minus Willie, they flew to Portugal. Here they began the European leg of their adventure, where among those they encountered, in Italy, was the explorer Thor Heyerdahl. Two years into their odyssey, walking through war-torn Afghanistan, both Dave and John were shot, after being ambushed by bandits. Dave survived but John, tragically, died of his injuries. At this point Dave took an extraordinary decision – to keep on walking. Assuming a kind of beatific countenance, one that led to him being dubbed the 'Earthwalker', Dave Kunst finally returned home in October 1974.

Among those inspired to follow in Kunst's footsteps (20 million in total, by Dave's reckoning) was the British adventurer and self-styled 'Runningman' Robert Garside. Setting out in 1997 after two abandoned efforts, he become a worldwide celebrity, completing his journey some six years later. It was only the start of his story; Garside ran in the early days of online posting, before the social media feeding frenzy brought instant reactions, and when claims made in error could linger in the digital ether, as doubters diligently checked your facts before trashing your reputation. Motivated to run by his desire to make history, that same desire, when it turned to scrutiny, came to haunt both Robert Garside and his fellow Briton Ffyona Campbell, who finished her own circumnavigation by foot in 1994. Campbell's infamous deception – accepting lifts during her walk – became headline news in 1996.

One whose reputation remains spotless is Romanian Dumitru Dan, the very first person to circumnavigate on foot (from 1910 to 1923), in a bid to promote the culture of his homeland. Originally part of a quartet, Dan was the only one to make it home, as three of his fellow walkers died en route. His efforts might not stand up to the scrutiny brought to bear on contemporary efforts, but there was no doubting his commitment; walking the entire route in traditional Romanian footwear, Dan is said to have worn out nearly 500 pairs of sandals.

# FOUR SET OUT – BUT ONLY ONE HEROIC WANDERER RETURNED

**Opium, gangrene and a fall killed three of the Romanians who pursued a cash prize (offered by a pioneering cyclist) for the first person to walk around the world.**

The first person to circumnavigate the world on foot owes his place in history to the humble bicycle. Founded in 1890, the Touring Club de France was an organisation that promoted the velocipede, an early form of bike whose various incarnations ranged from penny-farthings to bath chairs; the Club set great store by acts of exploration and endurance nonetheless. In 1908, under the auspices of its founder, Paul de Vivie, the Club announced a prize of 100,000 francs, to be awarded to the first person to complete a global walk of not less than 100,000km (62,135 miles).

One of the forefathers of French cycling, de Vivie was a champion of technology whose (often unfashionable) causes included the derailleur, the mechanism which enables cyclists to change gear while riding. He had proved the gadget's worth personally by cycling daily up the 1161m (3809ft) Col de la République, in the Loire, passing riders yet to adopt the high-tech kit. Tour de France founder Henri Desgrange dismissed de Vivie as appealing only to invalids and women; he was certainly known for his public support of female athletes, in particular the cyclist Marthe Hesse.

## WALKING ON WATER

It was in this spirit that de Vivie launched his walking competition, an egalitarian enterprise open to men and women; and in the knowledge that both sexes had acquitted themselves well during circumnavigations aboard his beloved velocipedes. The first man to cycle around the world, Thomas Stevens (see p.56), had completed a circumnavigation in 1886; the first woman, Annie Londonderry, came less than a decade later, in 1895. In the Paris of 1908, globetrotting races were in vogue.

In the end, the walking challenge was taken up by not one man but four – Romanian students living in Paris, George Negreanu, Paul Pârvu, Alexandru Pascu and Dumitru Dan. What followed

## ROAD FATALITIES

What befell Dumitru Dan's companions? In Bombay in 1911, one year into the trip, **Alexandru Pascu** died from an opium overdose. In 1912, **George Negreanu** fell while walking in China's Nan-ling Mountains and was killed. **Paul Pârvu** reached Florida before gangrene contracted in Alaska did for him.

was idiosyncratic to say the least. The four men returned to Romania to plan their trip. Their idea was to undertake a sort of cultural tour, seeking audiences with local dignitaries and spreading Romanian culture. Ahead of their departure, Dan and his fellow walkers embarked on several crash courses, in Romanian folk dancing, as well as lessons on the flute and accordion.

Accompanied by a dog, named Harap, the four set out from the Romanian capital, Bucharest, in late 1910. Dressed in traditional Romanian costume, they made first for St Petersburg, paying their way by performing folk concerts en route. From Russia, they walked to the Middle East, then on to Africa. With the Touring Club's stated minimum distance in mind, and concerned that the sea voyages on their expedition might be used against them – they might later be cited as periods of inactivity – the men spent many hours walking on deck, most notably on their voyage to Australia, where they had the ship's captain verify their distance.

## A BIZARRE ROUTE

What is known of their route tells of an erratic course. From Australia, for example, they sailed not for South America but turned back on themselves, to India; on the subcontinent, they walked to Siam (now Thailand), then sailed for Argentina. A major part of their walk across China came not after they visited Russia but Panama. And, having already visited Africa, they returned – and walked to Scotland. Over a period of six years, their walk around the world was not so much a circuit, as a crossing of every major continent, which is, interestingly, how many circumnavigations are plotted today.

Of the original quartet, only Dumitru Dan completed the route, his original companions having died over several years (see panel, left). Dan arrived home, in Bucharest, in 1916. WWI made the final section, a triumphant procession into Paris, impossible. He wasn't able to get there and collect his prize until 1923, 13 long years after it enticed him and his companions on to the road.

# THE RISE, FALL AND RISE OF ROBERT THE RUNNINGMAN

**Two decades ago, Robert Garside's claim to be the first man to run around the world crumbled under claims of inconsistencies and fabrication. End of story? Far from it.**

The story of Robert Garside, the first man to run around the world, is a complicated and surprising one. For one thing, it has no outright conclusion, at least not one that everyone can agree on. For another, prior to his remarkable six-year journey, Garside had almost no running experience; not for him the roster of extreme races and ultramarathons that garland the résumés of seasoned endurance athletes.

What makes Garside's story so compelling, however, is what happened once he'd hung up his running shoes. Doubts were raised over whether his round-the-world run claims were legitimate. A decade-long controversy followed, at its peak so unpleasant and personal as to send Garside into hiding. And then there came a heartening footnote: the willingness of one highly vocal critic to revisit a seemingly dead story and, in a commendable act of journalistic integrity, admit that an opinion to which he'd held fast for more than 10 years had been wrong.

## FINDING HIS WAY

Born in Stockport, Greater Manchester, in 1967, Robert Garside was far from a gifted athlete. At school he played football, but to no level of note. Having studied business at a Stockport college, he failed to complete a course – not once but twice. He dropped out again, from a degree course in psychology at Royal Holloway, part of the University of London. Various jobs followed; he served briefly in the Merchant Navy and volunteered with the City of London police.

Garside was inspired to run around the world by the man who had set the equivalent record for walking, America's Dave Kunst. His first crack at the running record was a low-key affair, one that began, and ended, in 1996. With no fanfare, he bought a one-way ticket to Cape Town, South Africa, and started running north, through Namibia. He planned to run along the entire west coast of Africa but aborted the attempt after just 1600km (1000 miles), realising he was woefully underprepared.

## THE ROAD BECKONS

Garside's second attempt, in December that year, was more organised. Setting out from London, he ran east, through Europe and into Asia. As before, he was entirely unsupported, although by now he had acquired various sponsors, among them the environmental charity Greenpeace. By September 1997 Garside had made it to Pakistan. The online journal he frequently updated reported harrowing experiences, narrow escapes that said much about his fortitude and determination. He had been shot at by locals in a remote part of Russia and, after managing to remain unscathed as he passed through war-torn Afghanistan, the incident-prone nature of his journey continued when he reached the Himalayas, where his luck ran out – he was robbed and his tent, his only refuge, destroyed.

Garside was forced to delay and, crucially, for more than 30 days, a fact that put him in breach of Guinness' rules (see overleaf); if he wanted to claim the record, he would have to start all over again. But by now, Garside's reputation for pluck preceded him. The many readers and news organisations who followed his website had been impressed when he declared that, following the attack in Pakistan, his worldly possessions ran to nothing more than the clothes on his back and his passport. And so, undaunted, on 20 October 1997, Garside began what was effectively his third attempt at the record, setting out from India Gate in New Delhi. His new plan wasn't simply to meet the criteria set by Guinness; this, broadly, required runners to complete a distance of no less than 37,000km (23,000 miles), starting and finishing in the same place and crossing every line of longitude. Instead, he aimed to set the bar so high his effort might never be beaten. To ensure his legacy, Garside's new route would take in the six continents required, not merely traversing them but doing so by the longest possible route.

## CONTINENT BY CONTINENT

From India he ran across Asia, to Shanghai in China, flying on to Japan, where he ran across Hokkaidō and Honshū before a flight to Perth, Western Australia. His route down under would, more or less, follow the coast all the way to Sydney. By the time he arrived at the town of Norseman in Western Australia, Garside had broken the record for the world's longest single run – 17,700km (11,000 miles) – held since 1988 by a US athlete, Sara Fulcher. Garside flew next to Chile, traversing South and Latin America by way of Brazil, Venezuela, Colombia and Mexico, before heading to San Francisco, turning right

Below: Robert Garside takes a break at the Sydney Opera House, Australia.

and crossing the North American continent.

He began to appear increasingly on local television, styling himself as 'the Runningman', a real-life version of the fictional Forrest Gump, and drawing others to run with him for parts of his route. From New York, he flew to Cape Town, running northeast from the Mother City, all the way to the border of Mozambique and Malawi. By now it was December 2001, three months after the September attacks on the World Trade Center.

With security being tightened around the world, Garside took the decision to fly from Mozambique to Morocco. From Rabat, he ran to Turkey, aware there were now gaps in his coast-to-coast crossing of Africa. In a bid to make amends – to run from one coast of the continent to the other – Garside arranged 'connecting' runs (effectively joining the dots) first in Egypt, and subsequently in Eritrea. Both came to nought, thwarted by the increasing political tensions in the region. Still determined to plug the holes, in April 2003 he flew back to Mozambique where, upon reaching Beira, he completed the African traverse that he began two years previously. Garside then took his final flight, to Kanyakumari, in the southern Indian state of Tamil Nadu, from where he made his way back to Delhi, returning to India Gate in June of that year. Nearly six years after he set out, and with over 64,000km (40,000 miles) of running in his legs, Garside's passport boasted stamps from 29 countries. His run seemed positively miraculous, given his inexperience and the many things that had gone wrong – in addition to the robbery in Pakistan, a brush with the authorities in China had led to Garside being jailed, on suspicion of spying.

## THE DOUBTS GROW

But among the wider ultrarunning community, the champagne was distinctly on ice. In 1998, with Garside a year into his third attempt, a Canadian journalist named David Blaikie had begun to cast doubts on the Runningman's achievements. Blaikie was an influential figure in the world of endurance running thanks to his hugely popular (now defunct) website, Ultramarathon World. Central to Blaikie's concerns were what he believed to be wild claims made by Garside about his time in Australia, in particular his crossing of the Nullarbor Plain, the 1300km (800 miles) of desert that extends from Norseman to Ceduna. Garside insisted he had crossed the inhospitable plain in four weeks, running unsupported and relying on such water as he could carry, or find during the journey. As challenging as the Nullarbor is, there are provisions for travellers; a series of 12 roadhouses are located, on average, around 96km (60 miles) apart, but sometimes separated

© ANDREW MURRAY / SHUTTERSTOCK

by as much as 190km (120 miles). Yet these are intended for drivers, as refuelling stops for cars; there are no natural water sources between them. Blaikie pointed out that Garside would have required up to 12 litres of water a day, a heavy load indeed.

Blaikie continued to monitor Garside's progress, compiling a dossier, mostly of runs he believed Garside could not have made in the times documented. Until 2000, however, this remained a relatively niche, if niggling, affair, an argument between extreme athletes and their record keepers. When, in May of that year, Garside chose to reply to Blaikie's criticism, the website editor's response was to publish the Englishman's splenetic emails online.

It was another nine months before the first blow to Garside's credibility was delivered by the press, a broadside from which he would never really recover. In February 2001, as Garside was crossing North America, Britain's *Sunday Express* newspaper ran a story claiming Garside's 1997 account of Himalayan bandits was fake. Other outlets fuelled the fire with suggestions that, at the time Garside claimed to be fending off thieves in Asia, he was in fact at home in England, at the behest of his then girlfriend.

Above: Robert Garside at Sydney's Bondi Beach.

There was no denying some elements of Garside's story unravelled as easily as a loose thread on a sweater. There were the witnesses – those who claimed to have seen Garside in person at a specific location when his online reporting suggested he was somewhere else entirely. Subsequent press coverage speculated he had never even been to Pakistan, nor Afghanistan. Among the running community, the latter claim seemed, given the fate of John Kunst in that country in 1972, a particularly egregious transgression.

## A PUBLIC HUMILIATION

The story is made all the more bizarre by Garside's own confessions; by late 2001, he was freely admitting to the British press that numerous online postings had been false, that a combination of previous failures and onerous demands from sponsors had put him under undue pressure, and that the level of scrutiny had astonished him. His humiliation was not only very public but also deeply personal; Garside was portrayed as a fraud and a lunatic. Past failures were called upon as testament to his delusion – the abandoned study courses, the patchy employment record.

Of the gaffes revealed, some were positively schoolboy-like in their naivety. There was the occasion, in December 1999, when Garside was photographed with one of the most notorious British fugitives of modern times, the late Ronnie Biggs. Biggs had been on the run – a fact not lost on the publicity-hungry Garside – since his involvement, in 1963, in the Great Train Robbery, one of the highest-profile crimes in Britain in the 20th century. Garside encountered Biggs, resident in Brazil since fleeing his homeland, on the famous Copacabana Beach, arguably one of the most public places in South America, at roughly the time he was running through the Amazon rainforest, according to his website. Or the time, in 2000, that Garside claimed to have run from Mexico City to the US border in just 10 days, a time that would have broken all records.

And, as Garside prepared to set out from California, he had announced he would be doing so in the company of two skateboarders, Tyler Buschmann and Justin Hawxhurst. The pair had teamed up in their own bid to become the first to skate across North America. When the trio fell out, the skaters were only too happy to confirm that Garside was not a man of his word.

And yet, despite all this, Garside kept running. As far as he was concerned, his three attempts on the record were entirely separate; the attempt that he abandoned in Africa; the fraudulent attempt that ended in September 1997; and the legitimate one he had begun a month later and concluded in 2003.

## THE WILDERNESS YEARS

The debate rumbled for four more years until, in an unexpected reversal, the authorities at Guinness concluded that, despite falsifying information on previous occasions, Garside had in fact constructed a complete circumnavigation of the Earth. Then Guinness went back on its position (see panel).

For the next six years the story went quiet, at least in the mainstream press. Garside was a cut-and-dried charlatan, a joke. If he appeared at all it was as an occasional pub quiz question, the runner who lied about circumnavigating the globe. The ultrarunning community was certainly not in the business of forgiving or forgetting. In a sport so governed by numbers, the controversy continued to simmer. Garside's story became a cautionary tale, dredged from the archives with every subsequent attempt; a warning to runners claiming to have matched, or surpassed, feats built on fiction.

Technology moved on. Critics who had once sent unsolicited emails to Garside's site now made hay on forum posts. His name continued to be trashed online until, driven to distraction, the Runningman all but disappeared.

## AN UNEXPECTED ALLY

A postscript to Garside's story is one the man himself certainly could not have expected.

Of those who had lined up to beat Garside with the proverbial stick, one now offered him an olive branch. In 2012, a writer named Dan Koeppel, a previously outspoken critic of Garside and a correspondent for *Runner's World* magazine, wondered if it was possible that he, indeed the running community in general, had made a mistake.

By the time the two men met – a rendezvous was arranged in London without confirmation the Runningman would even turn up – Koeppel faced an unforthcoming, unsurprisingly haunted figure. He found Garside to be likeable, nonetheless, if not entirely trustworthy; the runner did not, for example, try to explain away the tall tales he had published in 1997. Establishing enough of a relationship to borrow, and photocopy, Garside's log books, Koeppel began to piece together the story of what happened next and began to uncover a very different tale to the one the press, including himself, had originally reported.

## THE FIGHTBACK BEGINS

Koeppel concluded that critics such as David Blaikie had arrived at their conclusions by methods that were 'not illegitimate but... by no means authoritative'. Plotting Garside's route on a map was one thing but the reality out on the road was more nuanced. Koeppel called potential witnesses around the world and was surprised to find a number willing to vouch for the Briton. In addition to passport stamps that seemed to accord with Garside's account of his third record attempt, some of those who had seen him run in person had been highly impressed.

One, a local businessman named David Walker, met Garside in California. The two men ran together for 32km (20 miles), until the 6.2-minute kilometre (8.5-minute mile) pace set by Garside forced Walker to quit. Garside continued, without reducing his speed, for an additional 64km (40 miles). Having run well in excess of two standard marathons in a day, he woke the next morning to traverse the 678m (2225ft) San Marcos Pass, and at more or less the same pace. Other witnesses testified to seeing him running, fast and alone, in remote and challenging locations; one in South America, another in Tibet, China. These, remember, were the days before GPS tracking.

Koeppel also uncovered anecdotal evidence about just how crazy, at its height, the Runningman circus became. There were contributors to Blaikie's website, for example, who repeatedly challenged Garside to races along the route; devotees who would (➔ p.145)

(➔ p.145)

## THE GUINNESS REVERSAL

In 2007, Guinness announced that the record for being the first man to run around the world did indeed belong to Robert Garside. Then, when the *Guinness Book of World Records 2007* was published Garside was omitted, nor did his name appear in subsequent editions. Tired of the furore, Guinness claimed the record had, as a category, been rested.

## HYPONATREMIA

Drinking too much water or energy drink can reduce blood sodium, causing cells to burst. Potentially fatal.

## NAUSEA

Gels and energy drinks may suffice in a marathon, but at greater distances, the body requires – but struggles to absorb – 'real' food.

# ULTRARUNNING: THE PHYSICAL CHALLENGES

## JOINT INJURIES

About 40% of running injuries are of the knee, which is true of ultrarunning; stress fractures, from overtraining, are typical, too.

## BREATHING

It's suggested that long hours on trails bordered with pollen-bearing plants explains the increased incidence of breathing problems.

## BLISTERS

Varying terrain and a greater chance of stones, water and other irritants in the shoe, increases the risk of abrasion.

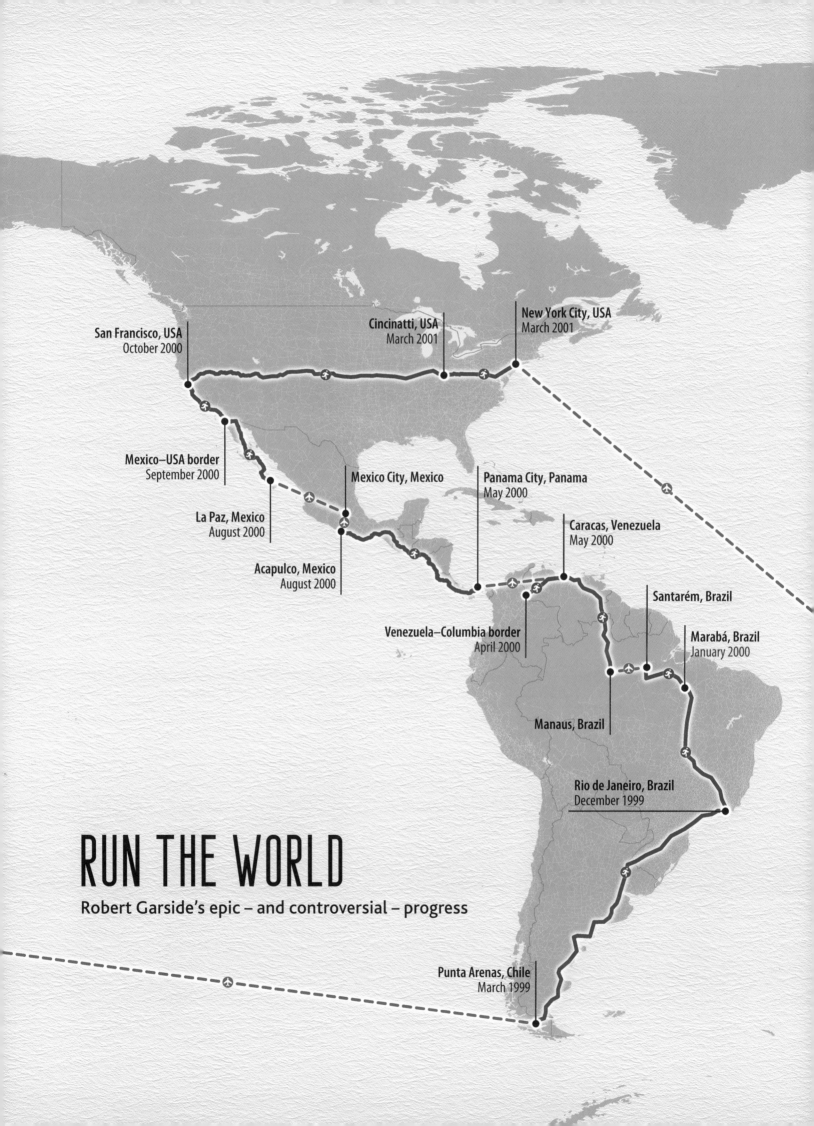

# RUN THE WORLD

Robert Garside's epic – and controversial – progress

**San Francisco, USA**
October 2000

**Cincinatti, USA**
March 2001

**New York City, USA**
March 2001

**Mexico–USA border**
September 2000

**Mexico City, Mexico**

**Panama City, Panama**
May 2000

**La Paz, Mexico**
August 2000

**Caracas, Venezuela**
May 2000

**Acapulco, Mexico**
August 2000

**Santarém, Brazil**

**Venezuela–Columbia border**
April 2000

**Marabá, Brazil**
January 2000

**Manaus, Brazil**

**Rio de Janeiro, Brazil**
December 1999

**Punta Arenas, Chile**
March 1999

# HOW DOES THE WORLD RUN?

The activity-sharing app Strava had 48 million members internationally in 2019 – here's how its runners shaped up

**21.8** BN
ELEVATION GAINED IN METRES...

**2.46** M
...WHICH IS EQUIVALENT TO THIS MANY ASCENTS OF MOUNT EVEREST

**7.6%**
COMPLETED MARATHON OR ULTRA RUN (USA)

**2721**
...WHICH IS EQUIVALENT TO THIS MANY RETURN TRIPS TO THE MOON

**2.09** BN
KM RUN...

**41:05**
MALE AVERAGE RUN DURATION

**5.9**
FEMALE AVERAGE DISTANCE IN KM PER RUN

**7.1**
MALE AVERAGE DISTANCE IN KM PER RUN

**39:11**
FEMALE AVERAGE RUN DURATION

SOURCE: STRAVA'S YEAR IN SPORT DATA REPORT 2019

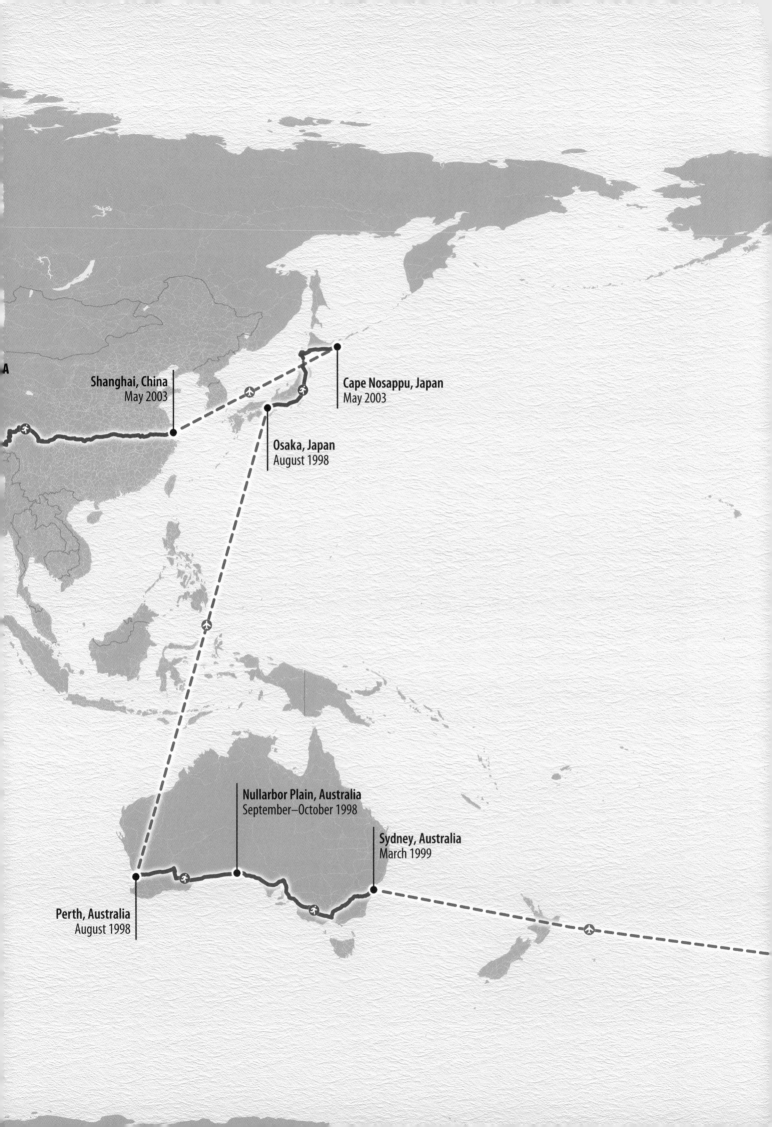

Shanghai, China
May 2003

Cape Nosappu, Japan
May 2003

Osaka, Japan
August 1998

Nullarbor Plain, Australia
September–October 1998

Sydney, Australia
March 1999

Perth, Australia
August 1998

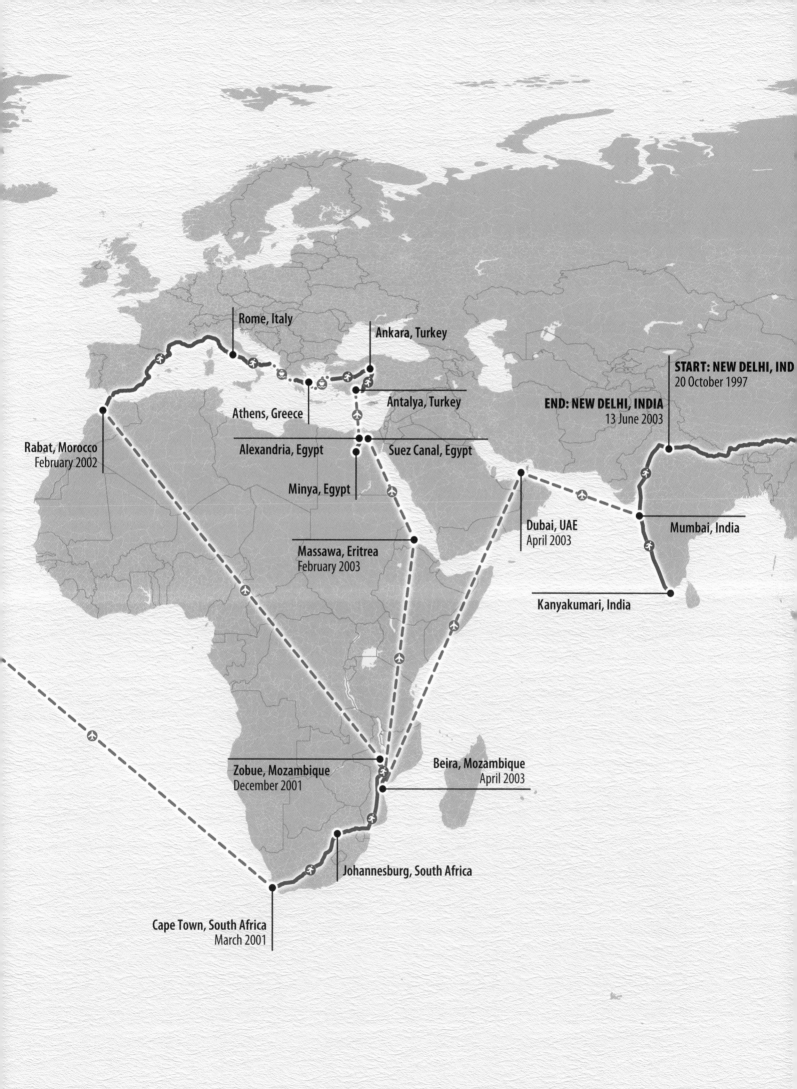

Rome, Italy

Ankara, Turkey

START: NEW DELHI, IND
20 October 1997

END: NEW DELHI, INDIA
13 June 2003

Athens, Greece

Antalya, Turkey

Alexandria, Egypt

Suez Canal, Egypt

Rabat, Morocco
February 2002

Minya, Egypt

Dubai, UAE
April 2003

Mumbai, India

Massawa, Eritrea
February 2003

Kanyakumari, India

Zobue, Mozambique
December 2001

Beira, Mozambique
April 2003

Johannesburg, South Africa

Cape Town, South Africa
March 2001

emerge from nowhere to try to hassle Garside into racing, or at least running faster.

Crucially, Koeppel found documents supporting some of Garside's seemingly most outlandish claims. His arrest in China, for example, was confirmed by official Chinese police papers, which Garside had secreted away with his notes and records.

There was proof of his competitive credentials, too. In 1994, Garside had competed in three marathons, two of which, in Brussels and Amsterdam, fell within 10 days of each other. In Brussels he had crossed the line in 41st place, a highly respectable position in a field of that class, and in the equally respectable time of two hours, 48 minutes.

And, in terms of ultrarunning, Garside's experience was not entirely without precedent.

## THE DEFENCE PUTS ITS CASE

There are tales – admittedly few – of endurance athletes who became proficient simply by the act of running and running some more. The American Dean Karnazes, for example, who on the occasion of his 30th birthday set off from his home and ran for 48km (30 miles) without pause, having not run for years. It's what you might call the 'Run-Forrest-run' principle. A little fanciful, perhaps – Karnazes is now a worldwide endurance-running celebrity – but not altogether impossible.

In the end, Koeppel says Garside gave up on protesting in the belief that whatever claims he made would always be used against him. When, for example he was invited, in 2003, to run 209km (130 miles) in 24 hours on a traditional 400m track – very different, of course, to his usual cross-country runs – his failure to do so was seized upon by his detractors. (The fact Garside quit after 14 hours, having clocked 115km/72 miles, went largely unreported.)

In 2013, Koeppel published a 9000-word article, again for *Runner's World* magazine. In it, he didn't so much absolve Garside as seek a better understanding of his motivation. If he was a fraud, Koeppel wondered, why had he stuck to his guns for so long? And, if Garside hadn't been running, what had he been doing for years on end? Koeppel suggested Garside might be the victim of his simple act of crying wolf; in being found out for things he hadn't done, he had removed any possibility of credibility in the perfectly legitimate runs he had made. Interestingly, as with Blaikie's original condemnation, at the heart of Koeppel's story lay Garside's account of his run across the Nullarbor Plain.

## THE NULLARBOR QUESTION

The points Koeppel sought to revisit – Garside's lack of support, the need for huge amounts of food and water – had been made many times, not least by Dave Kunst, the American who, during his own four-year global odyssey, had taken a mule into remoter locations; even being this relatively well-supplied, Kunst and his brother had still contracted amoebic dysentery.

A keen runner himself, Koeppel decided to return to the scene of the alleged crime – and attempt to run the across Nullarbor himself. What transpired seemed to prove his new theory; that running on the open road is very different to reading updates on a computer screen. And that what Blaikie could not have been aware of were the small, human interactions that might well have made Garside's run possible. Water, for example, was often donated by the drivers of passing cars. (During Koeppel's attempt, he was stopped by a driver who, having heard of his run, handed him a drink and some insect repellent.) If Garside found himself in a particularly desolate location, he would stop his clock, hitchhike to a motel or campsite, then return – in time permitted by the Guinness rules – to resume his journey.

When Koeppel made his Nullarbor run, he did so with a companion, Morgan Beeby. On a number of occasions, they found drivers willing to carry water supplies ahead, to be dropped at regular intervals. While such a method contradicted Garside's 'unsupported' status, it also showed unsupported doesn't always mean the same thing. 'The key to running the Nullarbor', Garside told Koeppel, 'turned out to be Australian hospitality' – not to be underestimated in such a challenging location. A local motel proprietor, Bob Bongiorno of the Balladonia Roadhouse, the first premises after Norseman, told how he and his staff followed Garside, collecting him from his finishing point and returning him to resume there, for several days consecutively. In the interim, Garside was given a bed and, in this way, claimed the Nullarbor was no more challenging than anywhere else. Certainly not the impossible monster of Blaikie's imagination.

Not that Koeppel didn't find the Nullarbor punishing. During the latter parts, he recalled, he felt as if he were running in a dream. The blisters he developed never fully recovered, reappearing years later if he ran more than a few miles.

## REPUTATION RESTORED?

The reality is we will never know for certain whether Robert Garside was the first man to run around the world. As far as Guinness is concerned, he's clearly tarnished goods. Dan Koeppel, who allowed Garside's story to occupy ten years of his career – by his own admission longer than it took Garside to circumnavigate the globe – maintains that he did.

# PATHS OF GLORY: CLASSIC HIKES AROUND THE WORLD

Can't manage a full lap of the planet? Whether you fancy 50km or 500km, we offer a quick guide to some of the best walks along the way.

## EUROPE

### SCOTTISH WATERSHED, SCOTLAND

The trail starts in the Cheviot Hills, at Peel Fell on the Anglo–Scottish border, and extends to the northeast tip of Scotland at Duncansby Head. Running for 1200km (745 miles), the Scottish Watershed is named for its close proximity to the 'border' of the Atlantic and North Sea. One for Munro-baggers – Scottish hills that stand higher than 914m (3000ft) – the trail is also dotted with 'Corbetts', whose own summits must be above 760m (2500ft). The Watershed is unmarked in places, so a GPS will most certainly help, and it can feel remote, but consider that a boon given how rare true wilderness is on mainland Britain.

### HAUTE RANDONNÉE PYRÉNÉENNE, FRANCE & SPAIN

Six weeks is the average time required to navigate this 800km (500-mile) trail. Zigzagging back and forth across the French–Spanish border, most of the route is located at high altitudes, so you'll need to be in good shape. Accommodation is provided by a network of mountain huts, although bear in mind these can get booked up and are sometimes located

Clockwise from top: the Scottish section of the Land's End to John O'Groats walk passes through loch-dotted Trossachs National Park, and the rugged Glen Coe; precipitous paths on the Haute Randonnée Pyrénéenne.

several days apart. So, you'll also need a tent, along with food and cooking gear. All of which tells you this is the real deal; tough at times but campers will be rewarded with wonderful morning views.

### LAND'S END TO JOHN O'GROATS, BRITAIN

Not one trail but rather a series, known collectively as the End to End; plan properly and you're guaranteed the best of Britain. The 1955km (1215-mile) route stretches from the far southwest corner of England, finishing in Caithness and the most northeasterly point of Scotland. You won't always be alone but that's part of the experience when tackling one of the most epic walks on Earth. An English departure will set you off on the South West Coast Path, taking in the Offa's Dyke Path and, eventually, the West Highland Way and Cape Wrath Trail, the last of which passes the tourist scrum of Glenfinnan (thanks for nothing, Harry Potter) but emerges in glorious Torridon.

### GR20, CORSICA, FRANCE

The GR20 crams in around 180km (112 miles) of high-altitude trekking, thanks to a snaking nature that closely follows the mountain ridges which form the island's spine (Corsica itself is only 214km/133 miles long). It follows that the trail is known locally as Fra Li Monti, 'Through the Mountains'. Choose to walk north to south and the route begins in Calenzana and ends in Conca; it's equally challenging in either direction. Highlights along the way include climbs of two peaks, Paglia Orba and Monte Cinto, but the standout is generally considered to be the Cirque de la Solitude, a vast rock basin requiring ladder access in places.

## NORTH AND SOUTH AMERICA

### PACIFIC NORTHWEST TRAIL, USA & CANADA

The PNT runs for 1930km (1200 miles), from the Continental Divide to the Pacific, starting in the Rockies in Montana and ending at Cape Alava, the westernmost point in the contiguous 48 states. In between, it skirts, but neve quite crosses, the Canadian border, eventually heading for the North Cascades. Thereafter, it skirts Puget Sound before climbing into the Olympic range. One of the US's newest trails, the PNT was conceived in the '70s, with the first guidebook published in 1984. Walkers can enjoy three national parks: Glacier, in Montana, as well as North Cascades and Olympic, both in Washington State.

### INCA TRAIL, PERU

Arguably the most popular walking destination in the world, of the Inca Trail's three principal routes, the toughest is the week-long, 72km (45-mile) Mollepata trail. Topping out at 4950m (16,240ft), its second half links up with another trail, known as the Classic. This covers a similar distance, of 80km (50 miles), from Cuzco to Machu Picchu, reaching the dreaded Dead Woman's Pass, at 4215m (13,828ft). Hikers normally complete the Classic in three to four days – its slightly easier route attracts the crowds, but the altitude means plenty still run, or rather walk, into trouble. A third, one-day trail is more popular still, stretching for 32km (20 miles) between the Sacred Valley and Machu Picchu. Highlights include spectacular views of the Andes and ancient Inca architecture.

### ARIZONA TRAIL, USA

Deserts present their own challenges for walkers, so the best (and most obvious) advice is to take plenty of water. The

Above: Corsica's GR20 takes you through stunning mountain landscapes such as the Plateau des Pozzi marshlands. Right: tackling Dead Woman's Pass on the Inca Trail, Peru.

© ALESSANDRO ZAPPALORTO/ SHUTTERSTOCK

Arizona Trail comprises 1300km (810 miles), from Mexico to Utah, encountering the 'sky islands' of southern Arizona along the way. The Chiricahua Mountains, to give them their proper name, are relatively small ranges at 2700m (8860ft) but are striking for the contrast between their forested flanks and the desert 'sea' below. Departing the islands for the Sonoran Desert, the trail then passes through two more ranges, the Mazatzals and the Superstitions. Its most dramatic feature is, of

Above: Olympic National Park on the Pacific Northwest Trail. Below: the Arizona Trail passes through the otherworldly Chiricahua National Monument. Opposite from top: Everest Base Camp; walking the Contintental Divide Trail at Grand Teton National Park, Wyoming.

course, the Grand Canyon, but the ponderosa forests of Utah, located on the Colorado Plateau, have their own majesty too.

### GREAT DIVIDE TRAIL, CANADA

One of the true wilderness trails, with much of the route both unmarked and remote, the Great Divide Trail (GDT) spans 1190km (740 miles) through the heart of the Canadian Rockies. It starts in Kakwa Provincial Park in Alberta and winds up in Waterton Lakes, near the US border. In total, it passes through five national parks; Jasper, Yoho, Kootenay, Banff and finally Waterton Lakes. The scenery is only ever jaw-dropping; soaring peaks cloaked in forests are home to moose and mountain goats. Grizzly bears and wolves top the food chain, and while undoubtedly a longed-for spot, they should be treated with extreme caution.

### CONTINENTAL DIVIDE TRAIL, USA

This 4989km (3100-mile) route meanders from the Canadian border down to New Mexico, taking in the Northern Rockies in Montana and, in Wyoming, the Great Divide Basin and the famed Yellowstone National Park. Wyoming is also the site of Two Ocean Pass, a rare hydrological feature that sends tributaries of North Two Ocean Creek to both the Atlantic and Pacific. The Colorado section of the trail boasts the Rockies, while New Mexico will tick boxes for desert-lovers. In places the CDT runs through open country, so sound navigational skills – or a GPS – are advisable.

## 03

## ASIA

### EVEREST BASE CAMP, NEPAL

A trek to the base of the world's highest mountain is an achievement in itself. Three-week hikes typically start in Lukla and you can choose your level of support. Depending on your route, sights may include the Tengboche Monastery, located beneath the 6800m (22,349ft) silhouette of Ama Dablam. As you draw closer to your destination, the views of Everest become as regular as they do spectacular. The sweeping Himalayan vistas

also include Everest's neighbouring peaks of Lhotse and Nuptse. The highest point of the trek is likely to be Kala Patthar, at 5644m (18,519ft) – Everest Base Camp is a lowly 5380m (17,600ft).

## MAKALU BASE CAMP, NEPAL–TIBET BORDER

At 8481m (27,825ft), Makalu may be only the fifth-highest of the world's 14 peaks above 8000m (26,247ft). But the 200km (125-mile) trek to its base camp is one of the toughest commercially available. The effort is well worth it; thanks to 7000m (22,965ft) of elevation, along the way you'll climb out of tropical rainforests to high mountain passes, which in turn deposit you on meadows that look up to glaciers beyond. The highlight of the trek, the Shipton La, is also its most testing section. The 4100m (13,450ft) pass offers sensational views, of the Barun Glacier in particular. The downside, if the altitude catches you out, is that it's also the only way back.

## AUSTRALIA

### OVERLAND TRACK, TASMANIA

Tasmania doesn't look like the rest of Australia. In fact, Tassie doesn't look like the rest of anywhere. The landscape here has an other-worldly quality and the Overland Track is one of the best ways to enjoy it. Most walkers give themselves between five days and one week to complete the 80km (50 miles) that connect Cradle Mountain and Lake St Clair, in the island's northwest. The Cradle Mountain-Lake St Clair National Park is a beautifully preserved wilderness in which the clearly marked, occasionally boarded path is never obtrusive. It passes lakes, forests and mountains, including the 1617m (5305ft) Mt Ossa.

**04**

Left: it's a tough trek up to Makalu Base Camp on the Nepal–Tibet border. Below: take the Overland Track to Cradle Mountain, Tasmania.

# ONCE YOU'VE WALKED THE WORLD, YOUR NEXT TEST IS TO PROVE IT

**Have you got the required minimum distance under your belt? Wandered across at least four continents? And the paperwork to back it up? Then welcome to the club.**

Those wishing to get a 'world walk' accredited can appeal to a variety of organisations to confirm a circumnavigation on foot. Guinness World Records will ratify a world walk, so long as it is a record of some kind (first, fastest, first on one leg, that sort of thing).

If you've nothing more impressive to offer than a plain plod around the planet, some admin is still required. Once upon a time, that meant dated passport stamps, proving arrival and departure; that, however, was all they proved. Journals and photographic evidence took things up a notch. A popular method among perambulating circumnavigators involved collecting the signatures of local dignitaries or officials. This – again, only in theory – confirmed you walked into town one day and walked out another.

Technology has, to a certain extent, come to the aid of world walkers; a GPS tracker can be checked on a regular basis, logging your route, the time taken and so on. However, even a GPS that accompanies someone travelling from London to Cape Town, for example, at roughly 6km/h (4mph), doesn't prove beyond any doubt that the distance was covered by walking.

To pick through the small print of any claimed circumnavigation, world walkers can submit evidence to their own governing body. Established in 2014, the World Runners' Association states that its mission is to '... officially represent legitimate runners and walkers who have, are, or are intending to circumnavigate the world in a way that is open, transparent and consistent with the WRA's rules and guidelines.' Said rules and guidelines are complicated – and in some cases seemingly contradictory – however, the basics are that walks must:

✪ Start and finish in the same place.
✪ Be a minimum distance of 26,232km (16,300 miles).
✪ Cross four continents coast-to-coast, walking a minimum of 3000km (1864 miles) on each continent.
✪ Cross all lines of longitude.

The WRA promises it will apply 'a rigorous yet unambiguous adjudication process' to the evidence of any attempted world walk, adding that, 'the overriding goal of the WRA is to ensure a consistency and uniformity in the rules for running and/or walking around the world, and honesty and transparency in any claims associated with attempts to do so.'

Glad we cleared that up.

## FOOT SOLDIERS

A few of the World Runners' Association current record holders:

Fastest world circumnavigation: Serge Girard, completed 8 April 2017, 434 days.

Longest circumnavigation: Tony Mangan, completed 26 October 2014, 50,000km (31,069 miles)

Youngest to circumnavigate: Jesper Olsen, completed 23 October 2005, 33 years 147 days.

# CROSS TRAINING: THE PREACHER WHO WENT ON A 69,000KM PILGRIMAGE

**Arthur Blessitt embarked on his first 'cross walk' in 1969 with such Christian zeal that Guinness World Records created a category all his own for his blistering exploits.**

The story seems almost fanciful. A US preacher who has walked around the world, on and off, for the best part of 50 years. A man called to do God's work who, in the process, has traipsed across every country on Earth. And who has done so while shouldering a life-size replica of the cross on which Jesus was crucified; all 18kg (40lb) of it. His name? Arthur Blessitt.

In 1969, Blessitt was a regular pastor, running a church-cum-coffee shop in the seedier part of Hollywood. And while it might be unfair to suggest he had a taste for the celebrity lifestyle, Blessit was regularly to be found preaching outside nightclubs. He would also set up his 'Jesus tent' at numerous rock festivals across the US. Those he encountered, in the dying days of the 1960s, included Janis Joplin and Jimi Hendrix. Blessit met the Beatles – and the Stones.

It was then that God spoke to him, requesting he remove the 3.5m (12ft) cross from the coffee-shop wall and carry it out into the world. Or at least to New York City, the destination for Blessitt's first 'cross walk', which he began on Christmas Day that year, covering a distance of 4500km (2800 miles).

The cross walks kept coming, as did the celebrities. Along with those you'd assume were well disposed to break bread with Blessitt – two Popes, the preacher Billy Graham – there are some unexpected names; Chief

Mangosuthu Buthelezi, for example, the Zulu statesman from South Africa; or the UK's Prince Michael of Kent. Blessitt's first overseas walk was in 1971, to a Northern Ireland in the grip of the Troubles. In addition to the great and the good, the other thing that seems to inspire him is conflict zones. During his half-century of perambulation, Blessitt has walked across Libya, where he met the late Colonel Muammar Gaddafi; and Palestine, where he drank tea with Yasser Arafat, then leader of the Palestinian Liberation Organization (PLO). He has also, for the record, carried his cross through North Korea, Somalia and Afghanistan.

It was in the Holy Land, in 1977, that Blessitt had, if not a crisis of faith, then at least cause to check in with the boss. Finding himself in Jerusalem, he appealed to God directly to ask if he should continue. The answer came, and then some. Blessitt says God instructed him to carry his burden across every country in the world, and to do so by the year 2000. Obeying

his orders to the letter, Arthur Blessitt has covered more than 69,200km (43,000 miles) in total, racking up visits to what he describes as around 320 countries, islands and territories.

In 2015, Guinness World Records acknowledged Blessitt's achievements by creating a category all his own, a record for the world's longest pilgrimage. Now in his early 80s, Blessitt finds he walks less and less in his home country, where he has been mistaken for a right-wing fundamentalist and even, on occasion, a hoodless Klansman.

Mostly, however, his cross walks are limited by plain old age. He continues to shoulder his burden but that old rugged cross isn't getting any lighter. In a rare concession to comfort, in the late 1990s he modified his cross, adding a wheel. Many are the wags who have pointed out that Jesus did not have the luxury of a wheel as he dragged his cross to Calvary. Blessitt's response is always the same: 'He didn't have to carry it as far.'

Above, from left: Arthur Blessitt with his family in 1971; Blessitt carries his cross around the lip of the Irazú Volcano in Costa Rica, 1978. Left: walking through Sydney, Australia, in 1976.

© FIONA HANSON - PA IMAGES / GETTY IMAGES

# THE FIRST WOMAN (NOT) TO WALK AROUND THE WORLD

The rewards for trailblazing world walkers are potentially enormous. But as the tale of Ffyona Campbell shows, so are the pressures – especially when life upends your plans.

**B**ritain's Ffyona Campbell was just 16 years old when, in 1983, she set out walking from John O'Groats, in the far northeast of Scotland. Bound for Land's End, the southerly tip of England, she reached her destination 49 days later, becoming the youngest person to complete the walk. This feat was to provide a spark for a far grander ambition – to become the first woman to walk around the world.

Planning to circumnavigate the Earth one continent at a time, Campbell's next odyssey

Above, from left: Ffyona Campbell in 1993, during the African section of her walk; cavorting on the beach in Tangiers. Right, from top: Campbell courted the press during her circumnavigation; camping in a WWII bunker in Calais.

was in 1985, when she walked across America, east to west. Unsurprisingly, it was tough going. Corporate sponsorship commitments meant long days of walking were rounded off with hours of interviews; it was on her walk across the States that Campbell laid the foundations for later infamy in the British tabloid press.

## FALLING BEHIND

Following a brief relationship with one of her support crew, Brian Noel, Campbell had become pregnant. As yet unaware of her condition, her pace slowed significantly. Soon enough, Campbell began to fall behind schedule, missing or cancelling appointments, including those organised by her main sponsor, the soup brand Campbell's.

One thousand miles into the walk, by now in Indiana, Campbell accepted the offer of a lift from Noel – a breach of the rules governing circumnavigations on foot. Tired, irritable and losing ground, Campbell took the first of a number of lifts estimated to have carried her as far as 1600km (1000 miles). When she reached New Mexico, she terminated the pregnancy, resumed walking, but said nothing to the press – or to her sponsor. She had begun to build a public following; her British walk alone had raised £25,000 for the Royal Marsden Hospital in London, the springboard for the £120,000 she would raise in total.

In 1988, at the age of 21, Campbell walked across Australia, from Sydney to Perth. She covered the 5150km (3200 miles) in 95 days, a record at the time. With two major continents under her belt, a place in the media spotlight was hers. In a time before social media, it is worth noting the extent to which Campbell became, and remained, a high-profile figure.

For more than a decade, Campbell appeared in newspapers and on television broadcasts across the globe. Her interaction with the press did not always run smoothly; the British tabloids made much of the fact she could be tetchy, portraying Campbell as a diva, a self-obsessed attention-seeker. It was all part of the circus.

In the early 1990s, Campbell spent two years walking across Africa before adding, in 1994, a walk through continental Europe. Starting in Algeciras, in southern Spain, Campbell walked to northern France, before crossing to Dover and returning, once more, to John O'Groats in Scotland. Eleven years after she had first set out, having plodded her way across a total of 31,520km (19,586 miles), Ffyona Campbell entered the record books as the first woman to have walked around the world. The lifts she had accepted in America, nearly a decade previously, went unremarked – for the time being.

## THE WHOLE STORY

In the public acclaim that followed, Campbell's abrasive nature might have become a distant memory, her spats with the press consigned to history. The then prime minister, John Major, even threw Campbell a reception at 10 Downing Street, praising her as a role model for young women. Yet the tag of drama queen was one Campbell could never quite shake. So, when her fall from grace eventually came, it seemed, to her detractors at least, inevitable. (Google 'Ffyona Campbell confesses' and one of the first results, from Britain's *Independent* newspaper, opens with: 'I never really liked Ffyona Campbell.')

Two years after completing her walks, in 1996, Campbell laid bare her deception in a book, *The*

*Whole Story*. Ever since her walk across America, Campbell revealed, her secret had gnawed away at her; every subsequent interview, each gala reception, only served to compound the lie. She even asked Guinness to remove her name from the *Book of Records*. Campbell's experience echoes that of Robert Garside, two stories that demonstrate why circumnavigations by walkers are often met with scepticism.

In *The Whole Story*, however, Campbell also details the potential consequences for those who undertake such epic challenges, the pressures they bring and what happens when things go awry; in Campbell's case, depression and a dalliance with heroin – the rock-bottom moment that persuaded her to come clean.

# CAR
## BY

F ew forms of transport have changed the world quite as comprehensively as the car. In barely a century, the internal combustion engine has opened up practically every corner of the earth to human exploration, transforming society. German engineer Karl Benz is generally acknowledged to have produced the first practical automobile in 1885, but it was American industrialist Henry Ford who gave the car to the masses with his pioneering Model T, which went into mass production in 1913. It was an invention that changed human history: from a few thousand cars at the turn of the 20th century to more than a billion at the start of the 21st, we all now live in the age of the automobile.

Unsurprisingly, it didn't take too long for drivers to start dreaming up adventures. The world's first long-distance road trip is said to have been made by Karl Benz' wife, Bertha, who in 1888 drove 106km (66 miles) from Mannheim to Pforzheim to visit her mother (the trip took just over six hours at an average speed of around 16km/h or 10mph). The next milestone came in 1903, when Horatio Nelson Jackson and Sewall K. Crocker became the first men to drive across the USA, taking 63½ days. The first woman, Alice Huyler Ramsey, repeated the journey six years later in 59 days.

From there, the next great driving prize was obvious: who would be the first person to travel around the globe by car? In a world where paved roads were still few and far between, and most cars managed only a few miles to the gallon, it seemed an almost impossible challenge. But the plucky young motorist Aloha Wanderwell pulled it off in 1927 – a globe-girdling feat that took more than five years. Aloha's record lasted barely a year; her great rival, the German racing driver, Clärenore Stinnes, completed her own two-year circumnavigation by car in 1929.

There have since been all kinds of madcap round-the-world road trips. Adventurer Garry Sowerby completed a second circumnavigation in a 4WD in just 21 days (smashing his previous time of 74 days). Serial circumnavigators Emil and Liliana Schmid have driven round the world more than six times in three decades. It's been done by bus, black cab, amphibious car and even by 'solartaxi' – the first circumnavigation powered by renewable energy, hopefully heralding a more ecofriendly future for the form of transport which, arguably, has done more damage to the planet than any other. Perhaps one day, we'll watch the first hydrogen- or fusion-fuelled car make its way around Earth. Maybe it will even drive itself.

To all those engaged in their own driving adventures – the grey nomads, snowbirds, Gumball Ralliers, digital dropouts and circumnavigating campervanners – we wish you empty asphalt and wide-open skies. After all, as the great poet of the American highway, Jack Kerouac, once wrote: 'the road is life'.

# THE CALL OF THE ROAD

From a weekend up Australia's Great Ocean Road to a half-year odyssey on the Pan-American Highway, via the Canadian Rockies, Namibia's Skeleton Coast and much else, get behind the wheel for the drive of a lifetime.

## SHORT-HAUL

### GREAT OCEAN ROAD, AUSTRALIA

**Length: 243km (151 miles)**
**Time required: 3–4 days**
**Start/finish: Torquay, Victoria/Allansford, Victoria**

Australia's vast size means that road trips here are often measured in weeks, not days – but the Great Ocean Road is one of the very few that you can cover in a weekend. Spinning past the beaches and cliffs of Victoria's coastline, the road was built in the early 1920s (prior to its construction, most of the coast was only accessible by rough bush tracks). These days, it's a popular jaunt from Melbourne, taking drivers through a varied landscape of seaside towns, world-class surfing breaks, sprawling farmland, lonely headlands and pockets of rainforest. Highlights include Cape Otway, a great place for koala-spotting, and the Twelve Apostles, a line-up of rock stacks that provides one of Australia's most iconic postcard scenes.

### SKELETON COAST, NAMIBIA

**Length: approximately 460km (285 miles)**
**Time required: 5–7 days**
**Start/finish: Swakopmund/Terrace Bay**

This ominously named road skirts the west coast of Namibia, an ill-famed graveyard for ships (wrecked vessels can still be seen along much of the route, along with the bleached bones of whales, seals and other sea creatures unlucky enough to find themselves beached on this arid coastline). You'll be passing mostly through dry, featureless desert, and are likely to encounter few other cars and even fewer signs of life (although the famous seal colony at Cape Cross is well worth a detour). That's assuming you can see the coast, of course: the thick fog that's foxed many a sea captain over the centuries can descend without warning. October to March generally promise the clearest skies.

### ICEFIELDS PARKWAY, CANADA

**Length: 230km (143 miles)**
**Time required: 1–2 days**
**Start/finish: Lake Louise, Alberta/Jasper, Alberta**

This glacier-studded road isn't just one of the most beautiful drives in Canada, it's one of the most beautiful anywhere on Earth. Locals modestly call it the World's Most Spectacular Drive, and it's not very hard to see why: the surrounding pristine wilderness comprises ice fields, glacial valleys, china-blue lakes, virgin forest and rugged peaks. Even better, there's a fairly good chance of spotting some local wildlife along the way: black bears, grizzlies, marmots and moose can often be spied moseying along beside – and occasionally on – the highway.

Clockwise from top: autumn colours along the Icefields Parkway in Alberta, Canada; Australia's Great Ocean Road offers eye-popping coastal views; a 4WD is essential in the dunes of Namibia's Skeleton Coast.

© FENG WEI PHOTOGRAPHY / GETTY IMAGES

© NIKKI BIDGOOD / GETTY IMAGES

## RING ROAD, ICELAND
**Length: 1332km (828 miles)**
**Time required: 7–10 days**
**Start/finish: Reykjavík**

The Ring Road (or Rte 1, as it's known to Icelanders) makes an unforgettable loop around the island's edge, passing some of its most stunning scenery en route: black lava fields, icy fjords, mighty glaciers, gushing geysers and smoking volcanoes that have a nasty habit of going pop without even the slightest of warnings. This is the land of the sagas, the sprawling medieval tales of Viking warriors, gods and great battles, and it's hard not to feel a little overawed by the drive. It can be done in just over a week, but it's worth allowing yourself at least two to factor in all the detours and photo-op stops. Summer is the most sensible time to tackle it: if you're mad enough to contemplate a winter journey, a 4WD equipped with snow chains is the very least you'll need.

## ROUTE DES GRANDES ALPES, FRANCE
**Length: 674km (418 miles)**
**Time required: 5–7 days**
**Start/finish: Thonon-les-Bains/Menton**

Infamous amongst cyclists and motorists alike, this incredible mountain road trip traverses the

# THE WORLD'S RISKIEST HIGHWAYS

### Yungas Road, Bolivia

It's only 69km (43 miles) long, but the road between La Paz and Coroico has been nicknamed 'the Road of Death' thanks to the extraordinary number of fatalities that have occurred along it. Unpaved, riddled with hairpin turns and barely 3m (10ft) wide at points, it also teeters over drops of almost 1000m (3280ft).

### Zojila Pass, India

Linking Ladakh and the Kashmir Valley, this perilous pass inches up the mountainsides to a dizzying 3528m (11,575ft), making it the second-highest road pass in the Himalayas. Though the pass is closed throughout winter, as soon as it opens hundreds of vehicles attempt it every day – not always with happy outcomes.

### Fairy Meadows Road, Pakistan

Edging up towards the world's ninth-highest peak, Nanga Parbat, this 16km (10-mile) road is a terrifying place to meet oncoming traffic: it's barely wider than a car, made of gravel and devoid of guard rails, despite the dizzying altitude. Nevertheless, it's surprisingly busy, and every year several vehicles plunge from it into oblivion.

### James Dalton Highway, Alaska

Often called the world's remotest road, Alaska's James Dalton Highway was originally built as a supply route for the Trans-Alaska Pipeline System in 1974. It's rough, gravelly, often icy and you'll mostly be sharing it with juggernauts. Even worse, you're an extremely long way from a breakdown assistance truck.

Above, from left: the Col du Galibier on the Route des Grandes Alpes is a challenge for cyclists and drivers alike; the Pan-American Highway cuts a swathe through Peru's Nazca Desert. Left: the coastal Ring Road takes in key Icelandic sights like Vatnajökull National Park. Right: juggernauts thunder along Alaska's James Dalton Hwy.

French Alps, taking in 16 of Europe's highest road passes along the way. Opened in 1913, the road wasn't fully asphalted until 1937, and even now it's only fully passable for around three to four months a year (usually June to October), when the annual snowpack has melted. The road's apex is the legendary Col de l'Iseran (2764m/9068ft), the highest paved pass in the French Alps – a name that strikes dread into the hearts of Tour de France riders. It has only featured in the race five times since 1947; in other years it's either been snowbound or too windy to be used.

# LONG DISTANCE

## ROUTE 66, USA
**Length: 3940km (2448 miles)**
**Time required: 3–4 weeks**
**Start/finish: Chicago/Santa Monica, Los Angeles**

Call it what you will – the Mother Road, the Main Street of America or the Will Rogers Highway – Route 66 is the queen of American road trips. One of the nation's oldest highways, completed in 1926, this epic stretch of asphalt has seen drivers of all persuasions: Dust Bowl refugees escaping from the Great Depression, beat poets in search of inspiration and freedom, migrants heading for California's promised land, silver-haired snowbirds living out their retirement driving round the nation in bus-sized RVs. The road's importance dwindled after the rise of the interstates in the 1950s and '60s, but it's been revived by a band of enthusiasts who are dedicated to preserving this slice of the USA's motoring heritage. Around 85% of the original road is still drivable; along the route, you'll cross eight states (Illinois, Missouri, Kansas, Oklahoma, Texas, New Mexico, Arizona and California) and three time zones.

## ACROSS THE RED CENTRE, AUSTRALIA
**Length: 3748km (2329 miles)**
**Time required: 1–2 weeks**
**Start/finish: Melbourne/Darwin**

If you've always dreamt of re-enacting a scene from one of the Mad Max movies, then here's the trip for you. Cutting straight across Australia's sun-baked centre, this is the ultimate Down Under route: a long, hot, dusty drive past red desert, ancient rock escarpments and remote outback towns. Along the way, you'll visit tiny mining settlements such as Coober Pedy and Marla, swing through Alice Springs to see the sacred outcrop of Uluru, and continue north past the Katherine River and Kakadu National Park to the tropical city of Darwin. It's not unusual to go

hours in the outback without passing another car, so it's sensible to pack plenty of water and supplies, fuel up whenever you spot a gas station – and practise changing a tyre before you set off. Oh, and keep an eye out for critters on the road – especially kangaroos.

## TRANS-SIBERIAN HIGHWAY, RUSSIA
**Length: approximately 11,000km (6830 miles)**
**Time required: 3–4 months**
**Start/finish: St Petersburg/Vladivostok**

Everyone's heard of the Trans-Siberian Railway, but did you know there's a road equivalent too? Comprised of a network of highways spanning Russia all the way from the Baltic Sea to the Sea of Japan, this continent-spanning road trip has only been fully paved since late 2015 (many stretches of the road were first built by prisoners of the gulags forced into service during their imprisonment). While most of the western sections are fairly good, the further east you travel, the rougher the conditions become: landslides, washouts and floods are all possibilities, so a sturdy all-terrain vehicle is likely to be a very good idea. The best time to attempt the drive is from June to September, since snow and ice makes much of the route impassable for several months of the year. Russia also has one of the world's worst road safety records – so take care.

## PAN-AMERICAN HIGHWAY
**Length: 30,000km (18,640 miles)**
**Time required: 6 months or longer**
**Start/finish: Prudhoe Bay, Alaska/Tierra del Fuego, Chile**

Crossing one America not epic enough for you? Then how about driving across the pair of them, from Alaska to the southern tip of South America at Tierra del Fuego. Encompassing in excess of 30,000km (18,640 miles) of driving (depending on the exact route you choose to take), the Pan-American Highway (PAH) has been described as 'the longest motorable road in the world'. From Alaska, it runs down the US west coast into Mexico, then along the spine of Central America until it hits a dead end at the notorious Darien Gap – a 95km (60-mile) swathe of rainforest between Colombia and Panama, patrolled by ruthless smugglers and drug gangs. There's no way across by road, so the only option is to hop over by plane or ferry, then resume the trip at Turbo, Colombia, before continuing south via Quito, Lima, Santiago and Buenos Aires, all the way to Ushuaia in Tierra del Fuego. At an average of 200km (125 miles) per day, it'll take you the best part of half a year. And that's before you even factor in the toilet breaks.

Clockwise from top: all US interstates are designated part of the Pan-American, so your road trip can take in iconic landscapes such as Utah's Monument Valley on Route 163; Mississippi's antique Chain of Rocks bridge was part of the original Route 66; 50 shades of red in the Australian Outback.

# 'WANTED: BRAINS, BEAUTY — AND BREECHES'

So ran the advert that 16-year-old Idris Welsh answered in 1922, turning her into 'Aloha Wanderwell', film star, director and first woman to drive around the world.

One day in 1922, a 16-year-old girl by the name of Idris Welsh was browsing the pages of the *Paris Herald* when she spotted a rather unusual advertisement. 'Brains, Beauty — and Breeches', the advertisement read. 'World Tour Offer For Lucky Young Woman... A good-looking, brainy young woman who is as clever a journalist as her appearance is attractive is wanted.'

Idris read on, rapt with interest. 'Moreover, she must foreswear skirts', the article continued, 'and incidentally marriage — for at least two years, and be prepared to "rough it" in Asia and Africa, and wherever else Capt Walter Wanderwell takes her...'

For the young Idris — who spent as much time as she possibly could daydreaming about escaping from the drudgery of life at her French convent school — it was as though her prayers had been answered. Dizzy with excitement, she begged her sceptical mother for permission to apply. Though Idris wasn't to know it, that fateful advert was to open the door to a fascinating lifetime of adventure — one which was ultimately to carry her more than 610,000km (380,000 miles), transform her into a fashion icon and film star, and seal her place in history as the first woman to drive the entire way round the world.

## THE 'INDEPENDENT TOMBOY'

Born on 13 October 1906 in Winnipeg, Canada, the daughter of Robert Welsh and Margaret Headley Hall, Idris Galcia Welsh was never destined for a conventional life. Describing herself as an 'independent tomboy' who was forever getting into scrapes, she loved to be the centre of attention whenever possible (impromptu ukulele concerts were one of her specialities).

She spent her early years on Vancouver Island, where she lived with her mother, her sister Margaret (often known as 'Miki') and stepfather Herbert Hall, a rancher and army reservist whom her mother had married in 1909 (the fate of her birth father Robert Welsh is unknown; Idris subsequently adopted her stepfather's surname).

The young Idris was as striking in looks as she was in character: blonde, long-legged, high-cheekboned and an imposing 1.8m (6ft) tall by her early teens. She was fascinated by adventure novels, and dreamt of becoming a movie star like her idol Mary Pickford — but like many young girls of her day, instead of embarking on the life of freedom and excitement she craved, she found herself packed off to boarding school in Europe to learn how to behave like a proper young lady and, all being well, attract a suitable husband.

The advert in the *Paris Herald* was to change Idris' future forever. The man who had placed it was 'Captain' Walter Wanderwell — in reality a Polish entrepreneur named Valerian Johannes Pieczynski, a man with a canny knack for self-promotion and a somewhat shady past that had seen him arrested as a German spy during WWI. Since 1919, Wanderwell — or 'Cap', as he preferred to be called — had been engaged in a madcap quest to drive around the world, competing against his then wife Nell in two specially adapted Model T Fords. The objective of the expedition was ostensibly to promote world peace, particularly the work of the newly formed League of Nations, via Wanderwell's newly founded organisation, dubbed Wawec ('Work Around the World Educational Club'); in reality, it was also a vehicle for Wanderwell to fund his taste for adventure and enhance his celebrity (not to mention a very handy bit of PR for the Ford Motor Company).

Billed as the 'Million Dollar Wager', the challenge combined elements of endurance race, road rally and travelogue. The aim was to visit as many countries as possible by car; the project financed itself with speaking engagements, PR events and promotional films shot and edited on the road. By 1922, however, the expedition had hit

Right: with her movie-star looks, the rechristened 'Aloha' became the face of Walter Wanderwell's round-the-world drive.

the rocks. The Captain and his wife had split up and the enterprise had ground to a halt.

## GOODBYE IDRIS, HELLO ALOHA

The moment he spied Idris, however, Capt Wanderwell knew he'd hit box office gold. He hired her on the spot, christening her with a suitably showy new stage name – Aloha Wanderwell. Nominally, Aloha was engaged as the expedition's secretary and French translator, but within a matter of months, she'd become the star of the show and the public face of the expedition – appearing in newsreels and promo films, giving interviews to the world's press, and increasingly shooting and starring in her own movies. Like everyone else on the expedition, her uniform was prescribed by the Captain – army tunics, military breeches, leather belts, riding boots and flying helmet. The outfit made an incongruous contrast to Aloha's fine cheekbones, glamorous looks and Mary Pickford curls, and seemed not a little scandalous in an age when fashions were changing but women were still often expected to dress demurely.

That first expedition carried Aloha through 43 countries on four continents (albeit conducted in stints, rather than as one continuous journey, as the promo films and her own autobiography seemed to claim). Aloha was to take charge of the second team, Unit 2, in a Model T Ford she nicknamed *Little Lizzie*.

## MEETING THE RED ARMY

The journey officially began in Nice in October 1922. From here, Aloha travelled through Italy, France, Spain, Belgium and Holland, traversing a landscape still scarred by WWI battlefields (her stepfather, Herbert, had been killed at Ypres in 1917). She got as far as Poland before, exhausted, she returned home to France for a break while the expedition steamed on across Romania, Turkey and Syria.

© THE RICHARD DIAMOND TRUST

© THE RICHARD DIAMOND TRUST

Aloha rejoined the caravan in Egypt, camping at the foot of the Sphinx next to the Pyramids of Giza, near Cairo. Next (having apparently sneaked into Mecca, the first non-Muslim woman ever to have done so, according to Aloha), she crossed the Sudanese desert, sailed from Yemen to Bombay, acquired a pet monkey called Chango, took a brief detour to the Khyber Pass (another possible first, according to Aloha), drove to Calcutta, sailed to Penang and then drove onwards across French Indochina to Singapore. Famously, they had to resort to using crushed bananas for differential grease, and occasionally substituted boiled-down elephant fat for engine oil.

Next came China, where Aloha visited the Temple of Heaven and the Great Wall, and was taught how to fire mortars by the Chinese Army, and afterward travelled on to Mongolia and Siberia, where she was made an Honorary Colonel by the Red Army for being 'the first demoiselle to pilot a motorcar to Siberia'. The expedition then sailed from Vladivostok to Japan, Hawaii, and finally to San Francisco, arriving back on US soil on 5 January 1925. Just a couple of

Above: Walter and Aloha Wanderwell – the two met in 1922 when Idris, as she was then, was just 16 years old. Walter was shot dead by an unknown assailant in 1932. Left: Aloha is charmed at the Taj Mahal.

170

Above: Walter 'Cap' Wanderwell with one of the expedition's adapted Ford cars – Aloha (left) named hers *Little Lizzie*, and donated it to Henry Ford after the circumnavigation (it was melted down for the war effort).

# HOW HENRY GAVE MOTORING TO THE MASSES

Thanks to the Ford Model T, millions got behind the wheel.

## AS LONG AS IT'S BLACK...

Between 1914 and 1926, the Model T is only available in black, because, it's said, the black paint dries faster. With a top speed of about 65km/h (40mph), it has two forward gears, one reverse (many drivers prefer to reverse uphill as that gear is more powerful). Its simple four-cyclinder engine is easy to repair.

1927

## THE END OF THE LINE

At its peak, the ever-more refined production line allows Ford to produce a Model T every 24 seconds. Over the 19 years of production, the Model T represents half the world's car output, selling nearly 17 million cars, the vast majority in the United States – where it transforms society.

## PILE 'EM HIGH

Within six years, Ford's new production line reduces the time taken to make a chassis from over 12 hours to 93 minutes.

## FOR THE MANY

Five years after he formed the Ford Motor Company in 1903, Henry Ford announces: 'I will build a motor car for the great multitude.'

1903

1914

1908-1927

## MOTORING FOR ALL

With huge economies of scale, Ford is able to make good on his promise of cheap motoring. Between 1908 and 1927, the cost of a Model T reduces from $950 to $290.

## THE ORIGINAL DISRUPTOR

Industry orthodoxy was to pay the workers little, price the product high. Ford prefers high volume, high efficiency, high wages – not least to make his workers potential customers.

## FIVE BUCKS A DAY

In 1914 Ford begins to pay workers a minimum wage of $5 a day (double the industry standard), reducing the working day from nine hours to eight and introducing three shifts.

list Republics
Harbin: Sep 1924
Szepinghai: Sep 1924
The expedition braves the Chinese civil war!
ll of China, Shan Hai Kwan
Pogranitznia
Aloha is named an honorary colonel in the Soviet army.
Vladivostok
Peking
n a
Tientsin: Aug 1924
1924
Shanghai
Tokyo
Mount Fuji
Osaka: Sep 1924
NORTH PACIFIC OCEAN
e 1924
with the US Army
World Flyers on their
mnavigation!
Hong Kong
Angkor Wat
Saigon
Philippine Sea
MICRONESIA
Honolulu
Hilo Bay
apore
Dutch East Indies
MELANESIA
POLYNESIA
SOUTH PACIFIC OCEAN
Australia
New Zealand

months after celebrating her 18th birthday, Aloha had officially travelled more than halfway around the world – an astonishing achievement in an age when most women had rarely even left their home patch, let alone their own country.

Despite her tender age, however, there was an unexpected and rather adult complication. Somewhat inevitably, Aloha and the Cap had fallen in love on their globetrotting journey. They got married in April, just two months after Cap's divorce from his first wife Nell.

The rest of 1925 was taken up with crossing the US from San Francisco to New York; along the way, Aloha found time to hobnob with her childhood idols Douglas Fairbanks and Mary Pickford in Hollywood, and also enjoyed a welcome parade in Detroit thanks to the industrialist who had made their journey possible: Henry Ford.

On 20 December 1925, Aloha gave birth to her daughter, Valri, but within a matter of months the expedition was back on the road – this time, for some reason, with an armoured tank in tow. Having traversed the States, they sailed from New York to Cape Town in October 1926 (a transatlantic passage which also involved an outbreak of typhoid). By the time they landed, Aloha knew she was already pregnant again; her son Nile was born six months later in Johannesburg, in April 1927.

But even two children couldn't slow down Aloha's wanderlust; within a month, she had packed the infants off to Australia with her mother, while she ventured onwards through Africa accompanied by her sister Miki, filming Masai warriors, tracing the source of the Nile and contracting malaria before sailing from Mombasa back to Europe by the start of 1928. Although she seems to have made no mention of it in her journals (a rather astonishing omission given her capacity for self-aggrandisement), sometime in 1927 while driving from Italy to Marseille, a full five years after beginning her incredible journey, Aloha officially became the first woman to drive all the way around the world. She was just 21.

## TO THE LOST CITY OF Z

The Wanderwells' first full-length movie, *With Car and Camera Around the World* (partly directed by Aloha) was completed in 1929. By then the couple were world-famous. Her record-breaking car, *Little Lizzie*, was donated to Henry Ford – but rather sadly, and without Aloha's knowledge, Ford later declared that the car lacked 'historical importance' and had it sent for scrap metal as part of the WWII munitions drive.

The inaugural Wanderwell expedition was followed by a second in 1930–31 to the rainforests of Mato Grosso in Brazil's Amazon Basin. This time, the aim was to search for the lost explorer Colonel Percy Harrison Fawcett, who had disappeared in 1925 while on a trek to find the legendary 'Lost City of Z'. During one landing by seaplane, Aloha recorded the first known contact with the Bororo tribe; footage of the encounter made its way into the 1934 film, *The River of Death*, their only non-silent film to receive a theatrical release (as well as subsequent films including *Last of the Bororos* and *Flight to the Stone Age*). The footage is now considered an important anthropological document: the original film is housed in the Smithsonian's archive.

After their return from South America, Aloha and the Captain purchased a yacht, the 30m (100ft) *Carma*, with ambitions to voyage to the South Seas and film their progress. Just before they were set to embark, in December 1932, the captain was shot dead on board by an unknown assailant. A member of the crew was charged but acquitted. The murder remains unsolved. A year later, Aloha remarried – this time to Walter Baker, a cameraman who (rather appropriately for the automobile-mad Aloha) also happened to own a petrol station.

She spent much of the 1930s touring the world with Baker, including voyages to India, Hawaii, Cambodia, Mexico and Australia. She released several more films and published an autobiography, *Call to Adventure!*, in 1939, before the outbreak of WWII brought an end to her travels. After the war, she continued to give lectures and screenings, but increasingly moved into journalism, curation and film archiving. She gave her final performance at the Natural History Museum in Los Angeles in 1982, and then went into well-earned retirement in Newport Beach, California. Walter Baker died in 1995; Aloha followed him a year later on 3 June. She was 89.

## PRODUCT PLACEMENT

As they travelled, the Wanderwells showed different edits of their film *With Car and Camera Around the World*: 'We would barter with local merchants and Ford dealers for gas and services in exchange for endorsing their products,' Aloha told a film magazine in 1982. 'It was a "finance as you go" expedition.'

Below: Aloha takes a press call with the local wildlife.

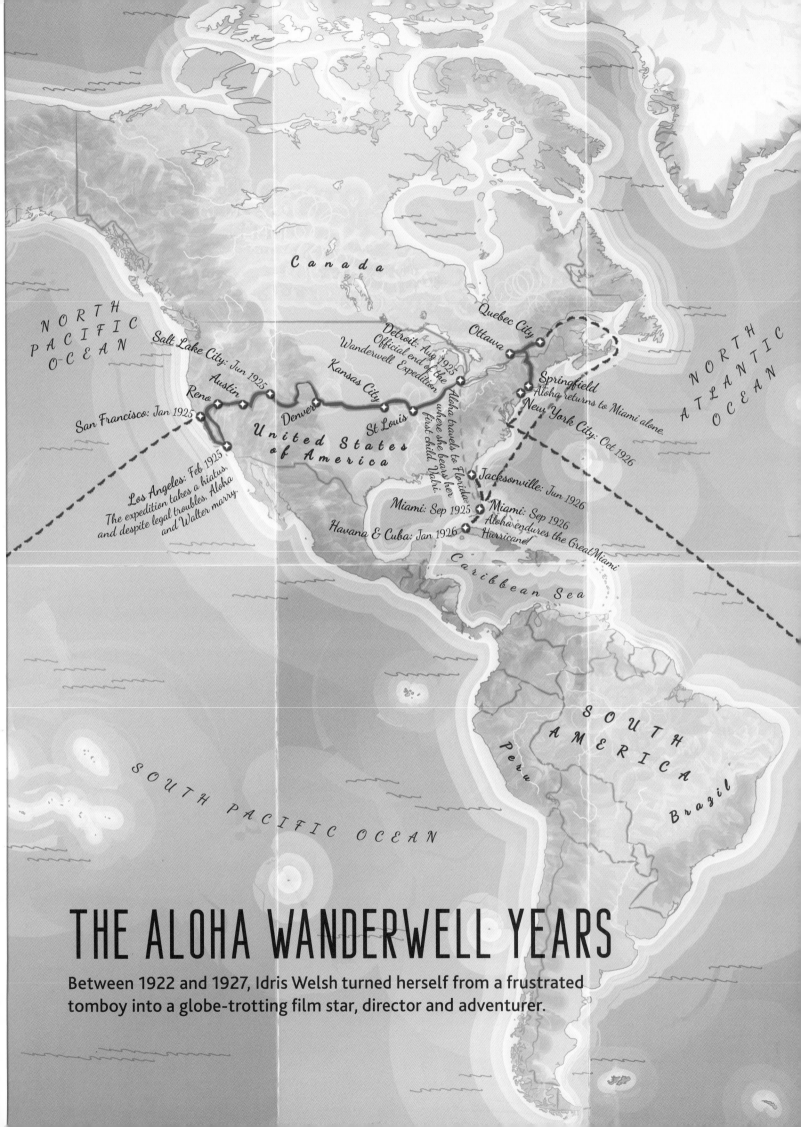

# THE ALOHA WANDERWELL YEARS

Between 1922 and 1927, Idris Welsh turned herself from a frustrated tomboy into a globe-trotting film star, director and adventurer.

Salt Lake City: Jun 1925

Austin

Reno

San Francisco: Jan 1925

Los Angeles: Feb 1925
The expedition takes a hiatus, and despite legal troubles, Aloha and Walter marry.

Kansas City

Denver

St Louis

Detroit: Aug 1925
Official end of the Wanderwell Expedition

Aloha travels to Florida where she bears her first child, Valri.

Quebec City

Ottawa

Springfield
Aloha returns to Miami alone.

New York City: Oct 1926

Jacksonville: Jun 1926

Miami: Sep 1925

Miami: Sep 1926
Aloha endures the Great Miami Hurricane!

Havana & Cuba: Jan 1926

United States of America

Canada

NORTH PACIFIC OCEAN

NORTH ATLANTIC OCEAN

Caribbean Sea

SOUTH PACIFIC OCEAN

SOUTH AMERICA

Peru

Brazil

Norwegian Sea

SCANDINAVIA

Union of Soviet Soci

Amsterdam: Aug 1923

Paris: Jun 1923

Berlin: Oct 1923

Warsaw: Nov 1923

Lemberg: Dec 1923

Start: Nice: Oct 1922
End: Nice: Dec 1927

Krakow: Dec 1923

EUROPE

The Carpathians
Snowbound and exhausted,
Aloha returns to Paris while the
expedition continues onwards...

Great U

Bilbao: Apr 1923

Oporto: Mar 1923

Madrid: Mar 1923

Rome: Nov 1922

Aloha crosses the Mediterranean to
rejoin the expedition in early 1924

C h i

Barcelona: Feb 1923

MIDDLE EAST

Khyber Pass

Cairo: Apr 1924

Karachi: May 1924

Agra & the Taj Mahal: M

Valley of the Kings

Mecca: Apr 1924
Aloha sneaks into the
holy site alone!

Benares

Calcutta: Ju
Aloha meets
Air Service
aerial circu

AFRICA

Aden: May 1924
Aloha forced to double back to
Massawa after passport problems

Aloha is the first woman
to drive across India!

Port Sudan: May 1924

Massawa: May 1924

Bombay: May 1924

India

Mongalla

Journey to Cairo halted by
strife in southern Sudan.

Ripon Falls
The source of the Nile!

Lake Victoria

Nairobi: Oct 1927

Mombasa: Dec 1927
Homeward bound to Paris
via Suez and Nice.

Georgetown, Penan
Aloha recovers after a bad ill
Sin

Mwanza

Dodoma

Mount Kilimanjaro

Messina: May 1927
The expedition is refused entry to
South Rhodesia on suspicion of
communist sympathies

INDIAN OCEAN

Umtali
Aloha is taken ill once more...

Johannesburg: Aug 1927

Pafuri: Jun 1927
A detour through Mozambique

SOUTH

ATLANTIC

OCEAN

Aloha travels to Durban
and Cape Town and bears
her second child, Nile.

Cape Town: Dec 1926

East London
Port Elizabeth

# THE FUTURE IS ELECTRIC

How better to demonstrate the advances in renewably powered vehicles than by racing around the world in them? No gas guzzlers allowed…

L et's face it: travelling round the globe by car isn't really the greenest option. In a world that's trying (and so far not succeeding) to wean itself off its century-old addiction to the gas guzzler, modern adventurers are setting their sights on more ecofriendly ways to propel themselves around the planet.

The first person to drive all the way around the world using only the sun's energy was a Swiss schoolteacher, Louis Palmer, who built his own 'solartaxi' with the help of several Swiss universities. He clocked up 54,000km (33,554 miles) with his co-driver Erik Schmitt, starting and finishing in Lucerne, and in 2009 was awarded the European Solar Prize for his achievement. Palmer then went on to establish the Zero Race, in which renewable-powered vehicles aim to race around the world in 80 days. The first edition was held in 2010, with entrants including a Swiss electric motorbike, an Australian three-wheeler and a modified VW Beetle from Canada; the race was won by Zerotracer, a two-wheeled electric vehicle built by a Swiss team especially for the event.

Although no one has circumnavigated in a wind-powered vehicle as yet, Germans Dirk Gion and Stefan Simmerer managed to clock up 5000km (3100 miles) in Australia, travelling from Perth to Melbourne between January and February 2011. Their lightweight car – a frankly astonishing 220kg (485lb) – was powered by a wind turbine charging a lithium-ion battery pack, with extra propulsion provided by a parasail (when the wind was blowing hard enough).

## ZEROTRACER VITAL STATISTICS

Capacity: 2
Length: 3.6m (11ft 10in)
Acceleration: 0-100km/h (62mph) in 4.25s
Top speed: 250km/h (155mph)
Range: 350km (217 miles)

Left: Louis Palmer's 'solartaxi'. Clockwise from bottom left: all angles of the Zerotracer, winner of the first Zero Race circumnavigation.

© GARRY SOWERBY

# DRIVEN BY DRIVING

Garry Sowerby has managed two global circumnavigations, by Volvo and Vauxhall: an end-to-end of the Americas and a spin from the tip of South Africa to northern Norway.

Few drivers are as globetrotting as Garry Sowerby. A former pilot and captain in the Canadian Armed Forces, Sowerby has spent the last four decades cooking up crazy driving challenges – from completing a circuit round the Eastern Bloc in a week to celebrate the fall of the Berlin Wall, to driving a displaced pelican thousands of miles back home after a hurricane, and from putting a car on top of Toronto's CN Tower to smuggling 5000 children's books into schools and libraries in Moscow. But he's best known for his circumnavigations: he has raced round the world by road not once, but twice.

Sowerby's first circumnavigation was in 1980 in a Volvo 245 DL Wagon named Red Cloud. With his navigator Ken Langley, he travelled a total distance of 43,030km (26,738

miles) around the globe, starting from Toronto, Canada, in 74 days, one hour and 11 minutes; this beat the previous fastest time set in 1976 by American racing driver Johnnie Parsons, who drove a Pontiac round the world in 102 days. Seventeen years later, Sowerby repeated the feat in a 1997 Vauxhall Frontera SUV, this time starting from Greenwich, England. The team travelled 29,522km (18,344 miles) in an incredible time of 21 days, two hours and 14 minutes.

Not content with that, Sowerby has also raced from the southern tip of Africa all the way to the top of Europe (Cape Agulhas, South Africa to Nordkapp, Norway), and from the end of South America to the end of North America (Tierra del Fuego, Argentina, to Prudhoe Bay, Alaska). He now runs his own organisation, Odyssey International Limited, which creates motoring events and advises others on epic driving challenges. He was inducted into the Maritime Motorsports Hall of Fame in 2012.

**Above: Garry Sowerby with his Volvo 245 DL Wagon, Red Cloud.**

# COMING UP ON THE OUTSIDE

Clärenore Stinnes bettered Aloha Wanderwell's achievement by becoming the first woman to make a true motor circumnavigation of the globe in 1929 (with three ballgowns).

Aloha Wanderwell may have been the first woman to drive all the way round the world, but she wasn't actually the first person to complete a true circumnavigation.

That claim belongs to German racing driver Clärenore Stinnes, who embarked on her own round-the-world drive in 1927. She set off from Frankfurt on 25 May 1927, finishing in Berlin (just past her starting point) over two years later, racking up a total of 46,063km (28,622 miles). She was accompanied by a Swedish film-maker, Carl-Axel Söderström, and two mechanics, along with a support vehicle loaded with spare parts, three ballgowns, a box of dynamite, cigarettes and 178 hard-boiled eggs. The vehicle she chose for her record-breaking journey was a mass-production Adler Standard 6 automobile, which

had only three gears and pulled a rather paltry 50HP (Clara's only concession to comfort was two lounge seats).

From Germany, they travelled through the Balkans and on to Moscow, via Beirut, Damascus, Baghdad and Tehran, then across Siberia and the frozen Lake Baikal. Next came an epic crossing of the Gobi Desert to Beijing, followed by ferries to Japan, Hawaii and Peru, a traversal of the Andes Mountains to Buenos Aires, then a journey up through Central America into the US (President Herbert Hoover himself gave them a special welcome in Washington, DC). From New York, they crossed the Atlantic to Le Havre, finally arriving in Berlin on 24 June 1929.

Like the Wanderwells, co-drivers Söderström and Stinnes subsequently married; they later moved to a farm in Sweden and had three children. Söderström died in 1976, aged 82; Stinnes lived on for another 14 years, passing away on 7 September 1990.

Above: the expedition's Adler Standard 6 fills up in California. Left: Ulaanbaatar, Mongolia, photographed by Carl-Axel Söderström in 1928. Opposite, from top: Clärenore Stinnes behind the wheel; crossing the Gobi Desert, Mongolia.

# THE NEVER-ENDING ROAD TRIP

It began in 1984 when they left Zurich for a North American road trip – 36 years, 186 countries and over 740,000km later, Emil and Liliana Schmid are still trucking.

D riving round the world once is probably more than enough for most ordinary people, but apparently not for Swiss couple Emil and Liliana Schmid. On 18 October 1984, they set off from their rented flat near Zürich in their Toyota Land Cruiser FJ60.

They've been driving pretty much non-stop ever since.

Originally, the couple didn't intend to take such an extended sabbatical from everyday life. Fifteen years after getting married in 1969, they decided to take a break from work to go on a round-the-world adventure. They both gave up their jobs (Emil was a software developer, Liliana a secretary) and in autumn 1984 hopped

on an Icelandair flight to New York. Their beloved two-year-old baby-blue Land Cruiser, was already loaded on to a container ship bound for Montréal.

The Schmids' initial plan was to take a year out to explore North and maybe a little of South America, and then probably head back home to Switzerland.

Somewhere along the way, they got a little sidetracked.

As of 4 April 2017, Emil and Liliana held the world record for the longest driven journey, travelling through 186 countries to notch up a remarkable tally of 741,065km (460,476 miles) – a distance roughly equivalent to driving round the world more than 18 times. Since then, their trusty Land Cruiser has been through its third major overhaul and, in Blumenau, Brazil, clocked

Above: Emil and Liliana Schmid with the Land Cruiser FJ60 that has transported them around the world since 1984.

Clockwise from top:
cruising through Chile's
Atacama Desert; buying
fresh flatbreads in
Azerbaijan; celebrating 20
years on the road in the
Caribbean island of
St Martin.

© EMIL & LILIANA SCHMID

## TRAVELS WITH 'OTTO'

The Schmids aren't the only long-distance road-trippers to hit the record books. German Gunther Holtorf currently holds the world record for the most countries visited (non-consecutively) by car.

Holtorf's wanderlust was apparently inspired by 30 years working for German airline Lufthansa, where he spent countless hours looking down on the world's roads.

Finally, in 1988, he jacked in his job to embark on his own atlas-spanning road trip. Piloting his much-loved German vehicle, a Mercedes-Benz G-wagon nicknamed Otto, Holtorf visited 180 countries between 18 December 1988 and 18 October 2014. He was accompanied nearly all the way by his partner, Christine, who sadly died of cancer in 2010; after she passed away, he was joined by the couple's son, Martin, and a close friend, Elke Dreweck.

Holtorf's journey came to an end in Belarus in 2014, by which point he was just 15 short of the full sweep of the world's total tally of 195 countries. He hasn't quite hung up his driving gloves yet – he still enjoys puttering around in his vintage Mercedes E-Class, which he bought in Indonesia in 1979. The record-breaking Otto, meanwhile, now occupies pride of place at the Mercedes-Benz Museum in Stuttgart, having been acquired for the princely sum of €450,000 ($497,570).

up its 777,777.7th kilometre (some 483,288 miles). So far, the Schmids reckon they've consumed 185,000 litres (48,870 gallons) of fuel, crossed 530 borders, had 187 flat tyres and spent more than 7000 days behind the wheel, at an average speed of just over 35km/h (22mph). They've also travelled through 23 of the world's 24 time zones and claim to have visited more countries by car than anyone else (the aforementioned 186), although another long-distance driver, Gunther Holtorf, actually holds the official record.

Now both in their mid-70s, Emil and Liliana do not seem remotely inclined to park up any time soon – despite the fact that their furniture is still stored in the same flat in Wallisellen, near Zürich, which they left more than three decades ago.

You can follow Emil and Liliana's continuing adventures – along with a comprehensive tally of records, figures and driving statistics – at their website, www.weltrekordreise.ch.

From top: winter in the Bolivian Altiplano; in 2016, the Land Cruiser got a major overhaul in Borneo – its third since the Schmids' epic road trip began in 1984.

© BERLIN / GETTY IMAGES

# OTHER RECORD-BREAKING JOURNEYS

### LONGEST JOURNEY BY MILK FLOAT

**Distance:** 1659km (1031 miles)
**Date:** 3 June to 20 September 2015
**Set by:** Paul Thompson (UK)
**Average speed:** 16km/h (10mph)
**Number of countries visited:** 1

### LONGEST JOURNEY BY DOUBLE-DECKER BUS

**Distance:** 87,367km (54,289 miles)
**Date:** 6 November 1988 to 3 December 1989
**Set by:** Hughie Thompson, John Weston and Richard Steel (UK)
**Number of countries visited:** 18
**Vehicle:** London Routemaster double-decker

### LONGEST JOURNEY BY COFFEE-POWERED CAR (CAR-PUCCINO)

**Distance:** 337km (209 miles)
**Date:** March 2010
**Set by:** Martin Bacon (UK)
**Number of countries visited:** 1
**Vehicle:** Volkswagen Scirocco
**Average coffee consumption:** 1.6km (1 mile) per 56 espressos

### LONGEST JOURNEY BY TRACTOR

**Distance:** 25,378km (15,769 miles)
**Date:** 8 May to 23 October 2016
**Set by:** Hubert Berger (Germany)
**Number of countries visited:** 36
**Vehicle:** 1970 Eicher Tiger II tractor

### LONGEST JOURNEY BY CAR IN ONE COUNTRY

**Distance:** 58,136km (36,124 miles)
**Date:** 11 July to 9 November 2016
**Set by:** Greg Cayea and Heather Thompson (USA)
**Number of countries visited:** 1 (USA)
**Vehicle:** Subaru Outback

### MOST COUNTRIES VISITED NON-STOP BY CAR

**Distance:** over 245,000km (152,000 miles)
**Date:** 1 January 1999 to 5 January 2002
**Set by:** Jim Rogers and Paige Parker (USA)
**Number of countries visited:** 111

### LONGEST JOURNEY IN A GOLF CART

**Distance:** 1665km (1035 miles)
**Date:** 14 to 22 January 2017
**Set by:** Morugadi Venkata Krishna Reddy, Morugadi Vanaja and Morugadi Vamshi Krishna Reddy (India)
**Number of countries visited:** 1

### LONGEST JOURNEY BY DIGGER

**Distance:** 5649km (3510 miles), from Brisbane to Canberra, Australia
**Date:** 11 January to 20 April 2019
**Set by:** Norman Bartie (Australia)
**Number of countries visited:** 1

### LONGEST JOURNEY BY A FIRE ENGINE

**Distance:** 50,957km (31,663 miles)
**Date:** 18 July 2010 to 10 April 2011
**Set by:** Stephen Moore (UK)
**Number of countries visited:** 1
**Vehicle:** Mercedes 1124 AF named Martha

### LONGEST JOURNEY BY TUK-TUK

**Distance:** 37,410km (23,245 miles)
**Date:** 8 February to 17 December 2005
**Set by:** Susi Bemsel and Daniel Snaider (Germany)
**Number of countries visited:** 21

# DOUBLE-DECKERS, TAXI CABS — AND CARS THAT FLOAT

**One round-the-world expedition was dreamt up in a pub, another was a bet (for a free pint of beer) and the final eccentric odyssey was abandoned by half the crew.**

## THE WORLD'S LONGEST CAB RIDE

We all know catching a cab can be an expensive way to travel, but how would you feel if you saw a six-figure sum on the meter at the end of your journey?

The record for the longest journey by taxi was set in 2011 by three intrepid – some might say eccentric – UK drivers: Paul Archer, Johno Ellison and Leigh Purnell. They set off from Covent Garden, London, on 17 February 2011 in a black cab (to be precise, a 1992 LTI Fairway FX4 London black taxi) which they bought from eBay for £1200 and christened Hannah.

The trio met as students at Aston University, in Birmingham, UK. Their initial intention (which was dreamt up, like all the best-laid British plans, over a pint in a pub) was to drive all the way from London to Sydney. The eventual route they took was designed, in the friends' own words, as 'the longest route ever, because taxi drivers always take the longest way around'. Accordingly, they drove right up to the Arctic Circle, then back across Europe via the Balkans, through Turkey and into Iraq, Iran, Pakistan and Central and Southeast Asia. The car was then shipped to Darwin, Australia, from where the team drove to Sydney and boarded another boat to San Francisco; then it was on to New York. With the cab airfreighted to Israel, the final leg involved travelling through the Baltics and Eastern Europe before crossing Germany, Luxembourg and France en route home to the UK.

They arrived back in Covent Garden just over a year later, on 11 May 2012, having broken two world records, for both the longest and the highest journey by taxi. Along the way, they racked up 69,716km (43,319 miles), passed through 50 countries and raised more than £20,000 for the Red Cross. They also went on to publish a book about the journey: *It's on the Meter: One Taxi, Three Mates and 43,000 Miles of Misadventures around the World*. The final fare? Thanks very much, guv – that'll be £79,006.80.

## GOOD IN THE WET

For any road-based circumnavigator, there are several large and rather wet obstacles that tend to hamper the route-planning process – namely the world's seas and oceans. But Australian Ben Carlin (1912–81) found his own ingenious solution to the problem: an amphibious vehicle.

In 1948, accompanied by his equally intrepid wife Elinore, he set off from Montréal, Canada in his chosen steed: a modified Ford GPA amphibious jeep which he named *Half-Safe*. After several failed attempts to cross the Atlantic (which saw the couple variously asphyxiated by exhaust fumes, stranded due to propeller failure and blown off course by Atlantic storms), they finally arrived in England on 24 August 1951, settling in Birmingham to raise cash for the next stage of their adventure. The journey resumed in 1954, crossing Europe to India, by which time Elinore had had enough and jetted back to North America. Carlin pressed on across Asia to the northern tip of Japan, chugging over to Alaska and finally back across Canada to Montréal.

His quest came to an end on 8 May 1958. In total, Carlin had covered 62,765km (39,000 miles) over land and 15,450km (9600 miles) by sea and river. He died on 7 March 1981. As yet, no one has repeated his epic amphibious journey.

## NEXT STOP, PLEASE, DRIVER

In 1969, in The Deers Hut pub in the little village of Liphook, Hampshire, nine young men were bet by their pub landlord, Bert Oram, that they couldn't get the whole way around the world by bus. Nothing inspires an Englishman like a completely ridiculous wager, so they accepted the challenge on the spot. A vehicle was duly acquired – a 1949 Leyland bus, which they bought for £100 with 700,000 miles (1,126,540km) on the odometer, and which they dubbed the Hairy Pillock 2.

The group set off in autumn the same year, progressing across Europe to Turkey and onwards to Iran. By the time they had reached Tehran, they were broke – so they did what any plucky young Brits would do when short of funds and formed a folk-singing group to raise some much-needed capital. Calling themselves The Philanderers, they quickly built up a local following; the Shah of Iran even attended one of their shows, although quite what he made of these nine young English eccentrics is sadly unknown.

Coffers restocked, the Hairy Pillock 2 forged ahead into Afghanistan, over the Khyber Pass, across India and Asia, eventually making it all the way to Australia, the US and Canada. By then, The Philanderers had become moderately famous; they were made honorary citizens of Texas, were given the keys to New York City and somehow even found time to make a record.

Two years and 10 months after leaving Liphook, they arrived back in England where they were greeted by Prime Minister Edward Heath, before returning to a triumphant welcome at The Deers Hut. Wager duly won, they claimed their prize: a free pint of beer each. Their marathon bus journey had covered 75,600km (47,000 miles) – a long way to go to get served, that's for sure.

Below: Ben Carlin atop his amphibious jeep *Half Safe* in Hong Kong Harbour, China.

# BY BALLOON

As the 21st century loomed, the annals of aviation featured one glaring omission – the absence of a circumnavigation of the world by balloon. When *Breitling Orbiter 3* became the first to do so, in 1999, it ended a decade of intense competition, during which those who vied for glory became household names, the likes of Per Lindstrand, Richard Branson and Steve Fossett. There was even a trophy, the Budweiser Cup, offered by the brewing company Anheuser-Busch to the first balloonists to circumnavigate the Earth without landing.

To focus minds there was also a cash prize, a cool $1 million. It seemed to do the trick; in the year leading up to *Orbiter's* successful circumnavigation, piloted by Brian Jones and Bertrand Piccard, no fewer than seven voyages tried and failed. Some were well funded, including one by then telecoms giant Cable & Wireless, while others were more modest, such as the solo effort of Kevin Uliassi, an architect from Arizona.

There are good reasons why this form of circumnavigation had taken so long. The very first balloon flight bore its aeronauts aloft as early as 1783, yet the balloon, designed by Jacques Charles, also returned said aeronauts (a duck, a rooster and a sheep) to earth 10 minutes later – in flames. Being suspended by a bag of highly volatile gas has long been the balloonist's dilemma. The first significant open-water crossing, over the English Channel, took place two years after that pioneering farmyard flight, in 1785, piloted by Jean-Pierre Blanchard and John Jeffries.

It would, however, be the best part of two centuries before the major oceans were crossed by balloon; *Double Eagle II*, under the command of Ben Abruzzo, made it across the Atlantic in 1978, while a successor, the Abruzzo-guided *Double Eagle V*, completed the first transpacific flight in 1981. In the time it took balloonists to graduate from channels to oceans, fellow aeronauts had made it to the edge of space. In 1960, US Air Force Captain Joe Kittinger ascended to an altitude of 102,800ft – before jumping out, on his way to a free-fall record that stood for the next half century. It was eventually broken, by Felix Baumgartner in 2012, and then superseded by Alan Eustace (of Google fame) two years later. The Canadian Steve Fossett, who appears on numerous occasions in this book, became the first person to pilot a balloon solo around the world, although he considered it the consolation prize to the achievement of Jones and Piccard.

*Breitling Orbiter 3* over the Swiss Alps in 1999, shortly after setting out on the world's first successful balloon circumnavigation.

# LEAVING ON A JET STREAM

Hostile airspace, 300km/h winds, freezing temperatures: the two-man crew of Breitling Orbiter 3 had to contend with them all to successfully balloon around the world in 1999.

A In Jules Verne's *Around the World in Eighty Days*, Phileas Fogg largely travels by rail and steam ship. But thanks, perhaps, to the 1956 film adaptation, in which David Niven as Fogg takes to a balloon called *La Coquette*, there is an abiding association between ballooning and world circumnavigation. What, one wonders, would Monsieur Gasse, the pilot of *La Coquette*, have made of *Breitling Orbiter 3*?

In the 1990s, a series of attempts at global circumnavigation by balloon revealed where nostalgia met the unforgiving winds of the jet stream. The result caught the popular imagination, which sponsors were alert to: those vying to make the first non-stop circumnavigation by balloon were hoping to win the Budweiser Cup, offered by the Anheuser-Busch brewing company, and a $1-million prize. But the teams were not only competing with one another and the elements — they were also high-altitude pawns in the deadly geopolitics at the end of the 20th century.

Left and below: the co-pilots of Breitling Orbiter 3 Bertrand Piccard and Brian Jones pose with the capsule that was to be their (very cold) home for 20 days.

© NICOLAS LE CORRE / GETTY IMAGES

© NICOLAS LE CORRE / GETTY IMAGES

## THE RACE TO BE FIRST

The watch company Breitling entrusted its investment to seasoned hands. Britain's Brian Jones was an experienced pilot with an unblemished record in the Royal Air Force. Accompanying him was a Swiss, Bertrand Piccard, who also had notable pedigree, being the grandson of Auguste Piccard, a pioneering Swiss balloonist of the 1930s (see p.195).

Piccard and Jones began their voyage on 1 March 1999, launching from Chateau-d'Œx in the Swiss Alps. In order to travel eastward, they first flew south to pick up the jet stream over North Africa. The journey officially began at the westernmost point the balloon reached – W 9° 27' longitude, in the skies over Mauritania – before bearing east.

So, come the 1990s, why was a circumnavigation by balloon regarded as such a feat of aviation? For one thing, it was one of the few firsts still up for grabs. For another, elite ballooning is so much more than drifting on the wind. Balloonists tend first to train as aeroplane pilots, as Jones had done, where they acquire the relevant knowledge, of physics and global weather systems, for example.

Beyond the ability to generate lift, balloons effectively fly without power, reliant on the winds that circle the planet. That does not mean they are entirely powerless to influence their fate, however. With fuel a finite resource, the key to any balloon circumnavigation is harnessing the power of the jet stream winds. These start at around 10,000ft and, travelling

west to east, can propel a balloon at speeds of up to 320km/h (200mph). The downside is their capriciousness; these high-powered winds change direction suddenly and a balloon crew can quickly find itself pushed off-course. To improve reliability, therefore, pilots are required to use winds at lower altitudes, which travel more slowly yet allow for greater control of direction. The trick is knowing when to shift between them.

There are other job requirements, too; such as living in a confined space – for weeks on end. *Breitling Orbiter 3* consisted of a 55m-high (180ft) balloon, whose volume, equivalent to seven Olympic-sized swimming pools, was filled with a mixture of propane and hot air.

Beneath it hung a 5.4m-long (17.7ft) gondola, made of Kevlar and carbon fibre, to which were strapped cannisters of highly volatile propane. Inside, in addition to navigation equipment, was a bed (for one) and a flushing toilet in the rear. The two men also found room for a rare edition of Guy de Maupassant's *A Life*; a signed gift from the author to his friend, one Jules Verne.

## GAMBLING WITH GADDAFI

It wasn't just the weather that might pose problems for the mission. Various nations were resistant to granting balloonists access to their airspace. In particular, the Libyan regime of Colonel Gaddafi had proved repeatedly hostile, a stance which, in 1997, led directly to the failure of a round-the-world attempt by Steve Fossett (see p.196).

## IN THE BLOOD

Bertrand Piccard's grandfather Auguste Piccard invented and in 1931 flew a pressurised balloon gondola to an altitude of 15,780ft (50,770m). (See also p.195.) He invented the deep-diving Bathyscaphe (in which his son Jacques descended to the Mariana Trench, a depth of 10,916m / 35,815ft). Piccard also became the inspiration for Hergé's Professor Calculus in the Tintin books.

China and Russia could also prove flies in the ointment. For their flight, Piccard and Jones were granted access to Chinese airspace but only if they flew below the 28th parallel north (28° degrees north of the Earth's equatorial plane). It paid to be cautious – the consequences of crossing sovereign airspace without permission were all too fresh in the memory. In 1995, an American balloon strayed illegally into Belarusian airspace and was shot down by a military helicopter, killing the aeronauts on board.

Piccard and Jones were cut an early break. When Libyan air traffic control proved unusually co-operative, they passed over that nation's airspace unhindered. Their subsequent route took them over Asia and the Pacific, then on through Central America. The 20 days spent inside *Orbiter*'s gondola proved cold, as well as cramped. The cabin was pressurised and heated but maintaining an ambient temperature of 15°C proved impossible, particularly at higher altitudes. On occasion, the mercury dipped low enough that Piccard and Jones had to break ice on their drinking water and defrost fragile electronic components.

### 'ORBITER, THIS IS MISSION CONTROL'

Fortunately, they had a high-tech ground crew to back them up. Today, a couple of decent smartphones would, between them, have sufficient global positioning technology to monitor the progress of a distant balloon. Twenty years ago, the hardware required to track the *Orbiter* balloon ran to several banks of desks at the team's Swiss operations centre. The latest (for the time) weather-monitoring systems also provided reams of data, allowing pilots to anticipate the best winds.

Though *Orbiter* was equipped with an autopilot function, one that could automatically maintain altitude by firing the gas burners, Piccard and Jones had resolved to man the controls continuously. Their solution was to divide each 24-hour block into thirds. For eight hours, one man would be the primary pilot; for a second shift, he would serve as co-pilot, then spend the final eight sleeping. Add to that the regular contact from mission control and Piccard and Jones could rarely have felt alone.

Certainly not when they flew over China, where authorities warned them off for flying too close to the 26th parallel; as close as 30km (20 miles) in some places. Piccard and Jones reached their start point over West Africa on 20 March, at a cruising altitude of 36,000ft (10,970m). It had been their intention to land in Egypt, close to the Pyramids, but *Orbiter* drifted on finally touching down some miles south.

Having established a landmark in ballooning history, the $1 million prize money was theirs.

Not that *Orbiter*'s crew were about to paint the town red. The expenses incurred in round-the-world ballooning are, traditionally, kept secret, in part to save face when things go wrong. The sport's practitioners, of course, know roughly how much it takes to fund a solo or two-man gondola, the cost of a propane cannister and how many you will need. (The total cost of the *Orbiter 3* mission was put at around $2 million.)

In the great balloon race of the 1990s, return on investment would be hard to calculate. As a mode of transport, ballooning is at once extreme and archaic. So, while Piccard and Jones' flight certainly caught the world's attention as a tale of derring-do, in Piccard's native Switzerland the response was positively old-fashioned; the Swiss Post Office issued commemorative stamps.

Below: *Orbiter 3* returns to earth in Egypt following its successful round-the-world flight in 1999.

© PA IMAGES / GETTY IMAGES

# BALLOON ACCIDENTS

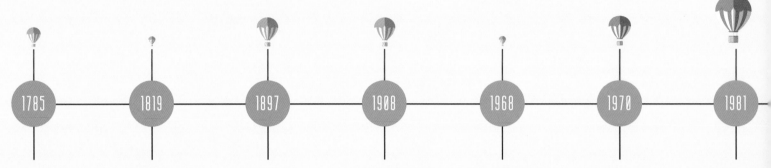

**1785** — **1819** — **1897** — **1908** — **1968** — **1970** — **1981**

### The English Channel
**1785**
Ballooning pioneer Jean-François Pilâtre de Rozier and a companion died in a crash attempting to cross the English Channel.

### Paris, France
**1819**
Professional aeronaut Sophie Blanchard died after trying to start a firework 300m over Paris, igniting her balloon.

### Kvitøya, Norway
**1897**
Swede S.A. Andrée and two companions mounted an expedition to fly over the North Pole. They disappeared; their remains were not found for 33 years.

### London, England
**1908**
At the Franco-British Exhibition, a balloon owned by American balloonist Captain Thomas Lovelace exploded, killing his young secretary and a man. Six others were injured.

### Vienna, Austria
**1968**
A balloonist, civil servant and journalist on board died when their balloon struck the Danube Tower, causing it to fall to the ground.

### North Atlantic
**1970**
Just 30 hours into its bid to fly the Atlantic Ocean, the *Free Life* balloon ditched in bad weather about 970 km southeast of Newfoundland, killing its 3 occupants.

### Illinois, USA
**1981**
A balloon hit power lines, causing it to ign and rise swift Five of six on board died.

# AIRSHIP CRASHES

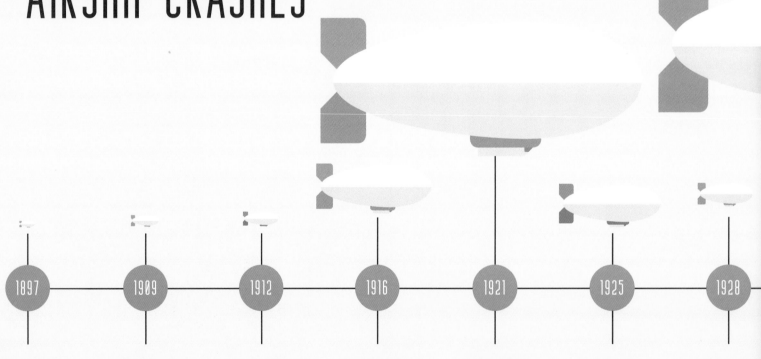

**1897** — **1909** — **1912** — **1916** — **1921** — **1925** — **1928**

### Berlin, Germany
**1897**
Friedrich Wölfert and his mechanic were killed when his petrol-powered airship *Deutschland* exploded at Tempelhof.

### Avrilly, France
**1909**
French Army airship *La République* crashed when a propeller pierced its envelope, killing all four crew.

### Atlantic City, USA
**1912**
Melvin Vaniman's second transatlantic balloon attempt ended when the *Akron* exploded just after take-off, killing all five crew.

### The North Sea
**1916**
After a raid on the UK, a German Zeppelin ditched in the North Sea. The 16 crew died when a British trawler refused to rescue them.

### Hull, England
**1921**
A British R38 class, then the world's biggest airship, failed structurally on a test flight. It crashed into the Humber Estuary, killing 44 of the 49 on board.

### Ohio, USA
**1925**
US Navy airship *Shenandoah* broke up in a violent updraft over Noble County, Ohio; there were 14 killed and 29 survivors.

### The Arctic
**1928**
The *Italia* crash returning from North Pole; sev died; a six-stro rescue team including explo Roald Amundse disappeared, a was never fou

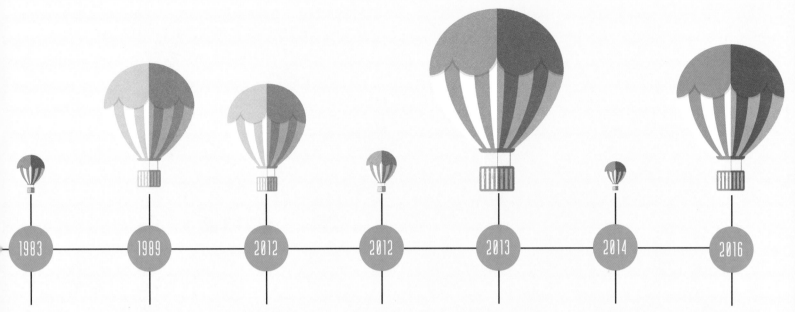

**1983**

**Albuquerque, USA**
After landing, a propane tank exploded, igniting the balloon gondola. As the balloon rose swiftly, four of the nine on board died either jumping or falling.

**1989**

**Alice Springs, Australia**
Thirteen died when one balloon collided with another, causing a canopy collapse.

**2012**

**Carterton, New Zealand**
Eleven died when a scenic hot air balloon hit power lines, ignited and crashed.

**2012**

**Ljubljana, Slovenia**
Wind shear caused a large balloon with 32 on board to crash and catch fire, killing four.

**2013**

**Luxor, Egypt**
A fire on a balloon flight killed 19 tourists when it crashed 300m to the ground. The world's worst ballooning accident.

**2014**

**Virginia, USA**
Approaching its landing, a balloon struck power lines, igniting it. The balloon rose swiftly and later crashed, killing three.

**2016**

**Texas, USA**
A balloon carrying a pilot and 15 passengers hit power lines and crashed, killing all on board.

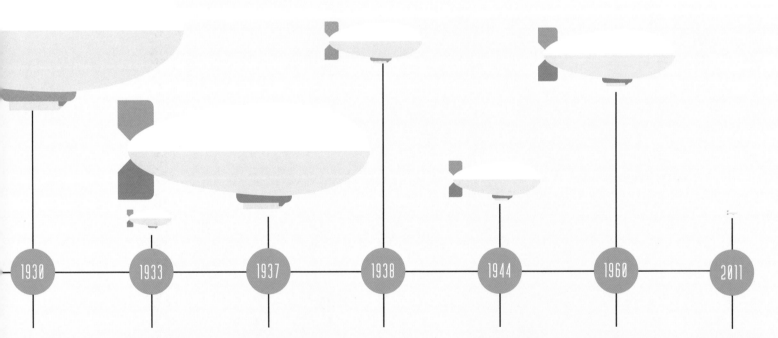

**1930**

**Allonne, France**
British airship R101 dived into the ground during a rainstorm; 48 died, six survived. The crash ended British airship development.

**1933**

**New Jersey, USA**
USS *Akron* was lost off the coast in a storm. Just three of its 76 airmen survived a disaster that signalled the end for the US Navy's rigid airships.

**1937**

**New Jersey, USA**
In the notorious disaster, German LZ 129 *Hindenburg* ignited landing at Lakehurst Though 35 died, remarkably there were 62 survivors.

**1938**

**Murmansk, Soviet Union**
En route to rescue members of a stranded Arctic expedition, the SSSR-V6 airship crashed into a hillside, killing 13 of its 19 crew.

**1944**

**Gulf of Mexico**
US Navy blimp K-133, operating from Houma, Louisiana, ditches in the gulf in a storm, killing 12 of its 13 crew.

**1960**

**Long Beach, USA**
A US Navy early-warning blimp crashed into the sea, killing 18 of 21 crew. The crash marked the end of the Navy's airship unit.

**2011**

**Reichelsheim, Germany**
A Goodyear Blimp ignited and crashed, killing pilot Michael Nerandzic, whose actions saved three passengers.

# FLOATING POINTS

**From flying sheep via aerial warfare to dare-devils plummeting from the edge of space, the history of balloon flight has been explosively eventful.**

The year 1783 was a busy one for French aeronauts. The first, unpiloted hydrogen balloon, designed by Jacques Charles, launched in August, while the first hot-air balloon took to the skies a month later, at the Palace of Versailles. Using convection from a small fire to fill its fabric canopy, the balloon's passengers were, allegedly, a duck, a rooster and a sheep. The flight lasted 10 minutes, until the balloon caught fire. The fate of the livestock was not recorded. The first piloted flight, in a hot-air balloon tethered to the ground, took place in mid-October 1783, manned by one Jean-François Pilâtre de Rozier, a physicist and chemist. (It had been the wish of the king,

Louis XVI, that the first pilots be convicts.) With a piloted hydrogen flight also taking place in September, the early science fell into two camps: the convection-filled balloons of brothers Joseph-Michel and Jacques-Étienne Montgolfier; and the hydrogen style pioneered by Charles. Worryingly, they weren't considered mutually exclusive.

## HYDROGEN BOMB

The first 'long-distance' flight came two years after de Rozier's historic voyage. Using a system which became common, a hydrogen balloon was tied to a basket, alongside a hot-air balloon. Unsurprisingly, the presence of a naked flame near a large quantity of volatile gas caused an explosion and de Rozier was killed just 30 minutes into the flight. The Frenchman's development was subsequently fine-tuned and used during a successful voyage, completed that same year, across the English Channel. The pilots on that occasion were another French balloonist, Jean-Pierre Blanchard, and American aviator John Jeffries. Buoyed by his cross-Channel success, Blanchard travelled to the US, where in 1793 he made a 45-minute flight from Philadelphia to Gloucester County in New Jersey. Large crowds gathered to see the first flight on the North American continent; among them was George Washington, the first US president.

## MILITARY MATTERS

By the mid-19th century, the military was exploring the potential of ballooning. In 1861, during the American Civil War, the Union founded a Balloon Corps. Under the Siege of Paris by Prussian forces (1870–71), balloons were used to carry people and supplies in and out of the French capital. In the 20th century, balloons

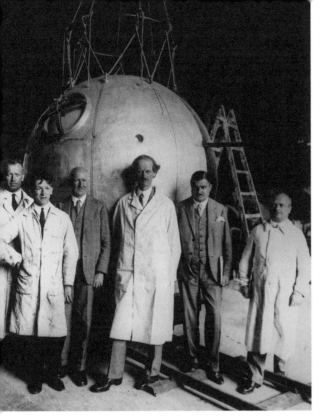

were employed throughout both world wars: to launch bombing raids in WWI; and as anti-aircraft cover in WWII.

## GOING STRATOSPHERIC

In May 1931, Swiss scientist Auguste Piccard, along with co-pilot Paul Kipfer, reached the stratosphere in a pressurised aluminium gondola. The pair set a new altitude record of 15,781m (51,775ft). Over the next few years, balloonists broke altitude records on almost a monthly basis. In 1935, a helium gas balloon, *Explorer II*, set an altitude record of 22,066m (72,395ft). Piloted by Albert Stevens and Orvil Anderson, the flight was a key milestone for future space flight, not least because of successful live radio broadcasts from the balloon.

## HIGH FLIGHTS AND MISTY VISORS

On 16 August 1960, US Air Force Captain Joe Kittinger jumped from a balloon at a breathtaking altitude of 31,333m (102,800ft). Kittinger's high-altitude jump and free-fall record stood for 52 years. It was eventually broken by Austria's Felix Baumgartner, in 2012, when he leapt from 38,969m (127,852ft). The octogenarian Kittinger acted as an adviser to Baumgartner; the two conferred moments before the jump when it was discovered that Baumgartner's visor fogged up as he exhaled (he jumped). His mark was broken just two years later, by America's Alan Eustace, from an altitude of 41,425m (135,908ft). Each was in free-fall for more than 4 minutes 20 seconds, reaching speeds in excess of 1285km/h (800mph). Observers on the ground heard sonic booms during both jumps. Kittinger later switched altitude for distance – in 1984 he flew 5690km (3535 miles), from Caribou in Maine to Savona in Italy. In recording the first solo

transatlantic flight, his helium-filled balloon was named *Rosie O'Grady's Balloon of Peace*, in honour of Betty Grable's character in the 1941 film.

## OCEANS APART

In 1978, *Double Eagle II* became the first balloon to cross the Atlantic Ocean; a successor, *Double Eagle V*, claimed the equivalent accolade for the Pacific just three years later, in 1981.

## FOR THE DURATION

In 1992, Richard Abruzzo, whose father Ben had piloted both *Eagle* balloons across their respective oceans, set a new record for the longest timed flight. Abruzzo, along with Troy Bradley, flew from Maine to Morocco in a time of 144 hours and 16 minutes.

Above, from left: Auguste Piccard (centre) and crew in front of their balloon gondola, 1930; Joe Kittinger makes his 1960 leap of faith from 102,800ft; Kittinger before the jump. Below: French military transport balloon during the Siege of Paris, 1871. Opposite: Felix Baumgartner reaches terra firma following his record-breaking 2012 free-fall.

# STEVE FOSSETT: ADDICTED TO ALTITUDE

**The serial adventurer spent millions of his own dollars over six attempts (one nearly fatal) before he finally completed his solo, non-stop circumnavigation in under 15 days.**

'**I**f at first you don't succeed', to quote Homer Simpson, 'give up'. The millionaire adventurer Steve Fossett was clearly not a regular follower of Springfield's first family. It was Fossett who observed that ballooning, the first form of aviation, was the last to complete a circumnavigation of the world. Why? The best part of a decade (and the small fortune) that Fossett spent trying to be the first to do it offers some clues.

Fossett was something of an adventure junkie. In 1985, he had swum the English Channel; in 1992, he competed in the 1450km (900-mile) Iditarod Trail Sled Dog Race that runs from Anchorage to Nome, in Alaska. Four years later, he drove in the 1996 edition of Le Mans, the 24-hour motor race. Dubbed 'Phileas Fossett', by 2004 he was the holder of 23 official world records for sailing. That same year he set a new, round-the-world sailing record, of 59 days and nine hours. Two years later, in 2006, he set one of many aviation benchmarks, this time for the longest flight, staying aloft for 77 hours.

One record, however, remained just out of reach: the first non-stop circumnavigation by balloon. In an echo of ballooning's early years, when 18th-century aviators crowded the skies, in the period around the turn of the 21st century the air was thick with record-breaking inflatables. A millionaire many times over thanks to his Larkspur Securities investment company, Fossett was certainly prepared to put his money where his mouth was. By the late 1990s, the scale of his financial largesse in pursuit of his ballooning dream was becoming the stuff of legend. In a seven-year period, he tried no fewer than six times to claim a first balloon circumnavigation, solo or otherwise.

The first attempt, in 1996, saw him fly just 2900km (1800 miles). A second, in 1997, estimated to have cost him $300,000, ended in frustration. Awaiting access to Libyan airspace – which ultimately wasn't granted – Fossett burned precious fuel, ran out of steam halfway and crash-landed in India. The 15,450km (9600 miles) he had flown at least set a different record, for both the longest duration and distance of a balloon flight, but it was scant consolation.

Another attempt, his third, in 1998, almost cost him his life. When a storm off the coast of Australia tore his balloon, Fossett had to endure a terrifying plunge into the Coral Sea, and one that began at over 8800m (29,000ft,) roughly the altitude of a cruising airliner. It was three days before he was picked by the Australian Navy, although he at least had the good grace to cover the $500,000 expense of the rescue personally.

While Fossett believed he would not survive a similar accident again, he nonetheless attempted to claim the record later that same year. Accompanied by British tycoon Richard Branson, along with Sweden's Per Lindstrand, their *Virgin Global Challenger* balloon made 13,315km (8275 miles) before they were forced to abandon, in Hawaii.

When Fossett was beaten to the first

## LAST FLIGHT
Steve Fossett's final flight, in Nevada in September 2007, was to be no more than a jaunt. He never returned; a huge search proved fruitless. In October 2008, his pilot's licence was found in remote terrain. It's thought a mountainous downdraft blew his 2-seater from the sky.

Clockwise from top: the Virgin Global Challenger at the start of the failed 1998 circumnavigation, which abandoned in the sea off Hawaii a week later; Fossett awaits departure of his 1996 solo attempt.

circumnavigation by balloon, in 1999 by Switzerland's Bertrand Piccard and Briton Brian Jones (see p.188), it was a bittersweet moment. Piccard and Jones were firm friends of the American, as was Branson.

In 2001, Fossett made a fifth attempt, in a bid to claim the only record available to him, a non-stop, solo circumnavigation of the planet. When that bid failed too, it allegedly cost him a staggering $1.25 million. So, when he resolved to make one final effort, even the deep-pocketed Fossett decided it prudent to seek commercial sponsorship. He was by now a celebrity, images of him ditching in some ocean or other a regular feature of news bulletins around the world. After securing backing (from brewers Anheuser-Busch), on 19 June 2002 he took off from Northam, in Western Australia, travelling west.

Fossett spent the next two weeks in a capsule measuring just 2m (6.5ft) long and 1.5m (5ft) wide. Living on tinned military rations and with only a bucket for a toilet, Fossett travelled more than 33,150km (20,600 miles). In reaching speeds in excess of 320km/h (200mph), faster than anyone had previously piloted a balloon, Fossett set yet another record.

His dream, if not his ultimate one, was realised when he passed Lake Yamma Yamma, in Queensland. Registering a total flight time of 14 days, 19 hours and 51 minutes, Fossett could at least toast his achievement with a beer, courtesy of his sponsors.

Three years later, in 2005, Fossett was at it again, when he became the first to fly non-stop around the world in an aircraft. The Virgin GlobalFlyer plane he used for the flight was sponsored in large part by Branson (see p.96), although the total costs for that venture were never made public. Fossett died in 2007, aged 63, when his plane crashed in Nevada.

Germany's Graf Zeppelin takes a test flight over Lake Constance in 1920.

# GRAF ZEPPELIN: THE HOTEL IN THE SKY

It was WWII rather than the Hindenburg disaster that ended the Zeppelin age. Until then, they were the epitome of glamour and luxury for global aeronauts and their passengers.

T he heyday of the Zeppelins occupies a unique place in aviation history. Billed as the ultimate in luxury travel, they reached their apogee in 1929, when the magisterial airship LZ-127, the 236m-long (774ft) *Graf Zeppelin*, made a landmark voyage; the first flight of any kind to provide passage around the world to paying customers.

And yet the political and historical climate in which these early airliners operated was far from straightforward – a decade or so after the war to end all wars, and 10 years before the war that proved otherwise. At their interwar peak, Germany's Zeppelin fleet offered confirmation of national resurgence, in engineering and design, but also in commerce, for these luxury liners were nothing less than the grandest of hotels in the sky.

Drifting around the world, seemingly carefree, for some they would never quite shake loose their darker associations. By the 1920s, the British public regarded Zeppelins with caution; they were, after all, aerial dreadnoughts (see panel, overleaf).

Towards the end of the 1920s, the *Graf Zeppelin* was, temporarily, the focus of a special relationship between Germany and America. It was a cosiness that would end thanks to events soon to play out – the crash on Wall Street and the increasing rise of Nazism.

What became the generic term for these slow-moving, aerial liners was in fact the family name of their creator, Count Ferdinand von Zeppelin. The first of his many prototypes to take to the skies, in the 1880s, were cumbersome and accident-prone, but by the turn of the century they had become as reliable as they were inspiring, soaring above the Earth in silent, regal splendour. Von Zeppelin's empire began in earnest with the Luftschiff Zeppelin 1 (LZ-1), whose maiden voyage, above Lake Constance, took place in July 1900. Barely a decade later, his aircraft were considered so reliable that von Zeppelin established an intercontinental airline, the first of its kind.

The most successful creation to roll off von Zeppelin's production line was the LZ-127, the only airship to bear the full family name, which it was awarded on 8 July 1928 (von Zeppelin had died 11 years previously, aged 78.) Its maiden flight was a few months later – and it was an aircraft that would capture the imagination of the world.

## DON'T CALL IT A BALLOON

In keeping with all the aircraft created by the Luftschiffbau Zeppelin company, founded in 1908, the LZ-127 was designed by Ludwig Dürr, an aeronautical engineer famed for his airship designs and who had been in the count's employ for the best part of three decades. A Zeppelin, of course, was not actually a balloon but a dirigible, meaning its canopy was supported by a solid, lightweight metal frame. Contained within were inflatable gas 'cells', usually made from cotton and coated with gelatine. Suspended beneath was the vast hull, which contained the passenger cabins, restaurants and recreation facilities, as well as viewing decks upon which passengers could take their daily promenade. The engines were located on the outside of the canopy and, by way of an open-air gantry, required constant

### DEATH FROM ABOVE

Zeppelins were pressed into military service during WWI. On 13 June 1917, a German Zeppelin carried out the first daylight aerial bombing raid on London. Among the casualties were pupils of Upper North Street School, in Poplar in the East End, killed when a bomb struck the building. Of the 18 fatalities, 15 were five-year-olds.

attention from those crew members tasked with their smooth running during a flight.

America's love affair with the *Graf Zeppelin* began almost as soon as she took to the air. On 15 October 1928, following a transatlantic crossing and subsequent tour – over Washington, Baltimore, Philadelphia and New York – the captain of the *Graf Zeppelin*, Hugo Eckener, docked his craft at the US naval station in Lakehurst, New Jersey. The airfield would become infamous for what transpired there nine years later, and the fiery fate that befell the *Graf Zeppelin*'s sister ship, the *Hindenburg*. But all that was to come. For now, the American public greeted the *Graf Zeppelin* with the utmost enthusiasm. The Atlantic crossing alone had made Eckener a celebrity and, following his landing at Lakehurst, he and his crew participated in a Broadway ticker-tape parade held in their honour.

## AN AMERICAN LIFT-OFF?

In some respects, the Graf Zeppelin's next feat, its round-the-world flight in 1929, bore similarities to the *Titanic*'s ill-fated transatlantic crossing 17 years prior. The vessel was a showcase for innovative technology and decked out to be conspicuously luxurious. Stamps were issued, commemorative memorabilia pressed into service. And, much like the *Titanic*, the *Graf Zeppelin* was making headlines long before its departure, its passenger list comprising, if not quite the great and the good, at least the well-connected. There was, for example, nobody of the standing of Benjamin Guggenheim, no John

From far left: Captain Hugo Eckener and crew before the *Graf Zeppelin*'s maiden transatlantic flight in 1928; Kaiser Wilhelm II with Count Ferdinand von Zeppelin, who pioneered the design of these iconic airships. Opposite: the *Graf* dwarfs spectators in Tokyo during the second leg of the circumnavigation.

Jacob Astor. Yet the *Graf*'s manifest still had its share of millionaires, minor aristocrats and prominent public figures.

Among the 60 passengers was the recently married, and later knighted, Hubert Wilkins, an Australian polar explorer renowned for his adventures in Antarctica. He was joined by Bill Leeds, the American millionaire who, at the tender age of 26, was heir to the 'tin plate' fortune built by his father, William Bateman Leeds. The US military was well represented too, in the shape of Charles Rosendahl, a staunch supporter of airship technology and later a Vice Admiral in the US Navy; and Jack Richardson, who joined Rosendahl as an official observer. They were accompanied, in turn, by diplomats, from the US and Germany but also from the Soviet Union and Japan. Medical assistance was provided by Dr Geronimo Megias, personal physician to Alfonso XIII, the king of Spain.

Of the *Graf*'s fervently excited press corps, the most prominent was Lady Grace Hay Drummond-Hay. The only female passenger – and who by default would become the first woman to circumnavigate the world by air – Drummond-Hay was a correspondent for the *Chicago Herald-Examiner*, one of the many titles owned by US newspaper tycoon William Randolph Hearst. It was Drummond-Hay's presence that led, in part, to disagreements about which flag the *Graf Zeppelin*'s world voyage should be made beneath. Hearst, as the principal backer, had met half the costs of the trip in return for exclusive media rights in the US and Britain, and was adamant the flight should start and finish in America.

His German counterparts thought differently. In 1929, the country was struggling to meet its post-war reparations, the hyperinflation of the early 1920s having heaped ruin upon catastrophe – the Weimar Republic would be

Below: the *Graf Zeppelin* flies over Berlin's Brandenburg Gate. Opposite: celebrated designer Fritz August Breuhaus de Groot created the *Graf*'s Bauhaus-inspired interior decor.

© ULLSTEIN BILD DTL / GETTY IMAGES

toppled just four years later. The idea of an American flag was an affront to the newly emboldened German industry (which seemed to occupy a parallel universe to the crisis-strewn path followed by mainstream German society). Naturally, the Zeppelin's owner-operators were certain any round-the-world flight would commence and conclude in Germany.

The solution was a compromise. There would be two official flights: an American version, which would begin and end at Lakehurst; and a concurrent German world flight, whose start and finish would be Friedrichshafen. In total, the voyage would cover 34,200km (21,250 miles), broken across five legs: from Lakehurst to Friedrichshafen; from Friedrichshafen to Tokyo; Tokyo to Los Angeles; Los Angeles to Lakehurst (completing the American flight); and, finally, from Lakehurst to Friedrichshafen, where the German flight would end.

The days leading up to the *Graf*'s New Jersey departure saw something of a social whirl, with receptions, dinners and even a 'Zep Ball', debutante-style, its guests comprised of passengers and celebrity well-wishers.

Keeping everybody happy led to occasional moments of farce. In order to commence its American flight, on 27 July, the *Graf Zeppelin* flew 7000km (4350 miles), from the German start point of Friedrichshafen to Lakehurst. It then made the return journey, a week or so later, whereupon the official German edition got underway, on 7 August.

## THE FLYING BAUHAUS

So what lay in store for those walking up the gangplank of the LZ-127? If not quite the opulent fixtures and fittings that would grace its successor, the *Hindenburg*, certainly a range of interiors befitting one of the great architects of the age. Fritz August Breuhaus de Groot was *Graf*'s go-to designer, his reputation burnished by the sleek interiors he created, in the Bauhaus style, for German film stars and celebrities. If the *Hindenburg* was his masterpiece, the *Graf* was his test bed; it was aboard the *Graf* that he honed his ideas, with large, well-dressed public spaces, the walls draped in rich fabrics, such as silk and light velvet. A primary concern was keeping weight to a minimum, so fittings were of plastic or aluminium, materials suited to the Bauhaus aesthetic. The two viewing decks were things of wonder too, allowing passengers to promenade untroubled by the weather. A great sweep of windows looked out on the sky and land below; above them, the belly of the taut, luminous balloon – an aluminium coating to protect it from solar radiation made the canopy glow at night – made for quite a ceiling.

The *Graf Zeppelin* was practical, too, capable of carrying huge loads, including motor vehicles. It was also perfect for transatlantic mail runs, albeit slow ones. For all their benefits, however, a distinguishing characteristic of Zeppelins was their vulnerability to extreme weather. Indeed, the *Graf Zeppelin*'s career had almost ended before it began when, during its first transatlantic crossing, in 1928, it encountered heavy storms. As a captain, Eckener was considered the safest pair of hands, with an uncanny ability to anticipate changing conditions. On this occasion, he flew into the oncoming front unusually quickly, causing the ship to rise suddenly.

Climbing too quickly, Eckener knew, was likely to result in the aircraft breaking up. Keeping an airship stable is a balancing act. Helium gas, being lighter than air, naturally creates lift. However, because the ambient pressure diminishes as the ship rises, the gas is continually expanding. Let it expand too much, and the gas cells would exert sufficient pressure on the airship's structure for it to be fatally compromised. To avoid over-expansion of the gas cells at higher altitudes, airships were fitted with automatic valves, designed to release helium as the ship gained altitude. These were watched, and in emergencies overridden by, the elevatorman, the crew member responsible for monitoring the pressure at which gas is stored inside the ship's canopy.

Crucially, it meant all Zeppelins were subject to a maximum rate of climb, beyond which newly expanded gas could not be released quickly enough. Eckener was all too aware of the fate of an American airship, the

## LIVING DANGEROUSLY

Remarkably, passengers aboard the Hindenburg enjoyed a smoking lounge mere feet away from nearly 200,000 cubic metres of hydrogen. Cigarettes, cigars and lighters were sold on-board. The lounge was monitored by the crew, who saw to it that all lights were extinguished before passengers left the special, pressurised cabin.

USS *Shenandoah*, four years previously. On 3 September 1925, the *Shenandoah* had been flying over Ohio when she was caught in a storm. Strong winds caused the ship to rise suddenly, eventually at a rate in excess of 1000ft per minute. The *Shenandoah's* valves had recently been adapted, thanks to some ill-advised tinkering by her captain, Zachary Lansdowne, for a rate of climb no greater than 400ft per minute. When the *Shenandoah* reached an altitude of 6000ft, and in a just a couple of minutes, the aircraft suffered catastrophic structural failure and broke into pieces. (Lansdowne had, in a bid to save weight, ordered the removal of 10 of the *Shenandoah's* 18 valves, modifications even the most vigilant elevatorman would have been unable to counteract.) Under Eckener's command, the *Graf Zeppelin* eventually made it safely through the storm and on to New York that day in 1928, although extensive repairs were required.

## TO TOKYO, AND BEYOND

One year later and in the public spotlight, Eckener was wary. The most challenging of the five legs facing the *Graf Zeppelin* was the second, and longest, phase of the journey, the 11,247km (6988 miles) that separated Friedrichshafen from Tokyo. The journey time was a staggering 101 hours and 49 minutes, four days aloft traversing much of Europe and the entire continent of Asia. The fact of Hearst's coverage made a visit from the *Graf Zeppelin* extremely prestigious, and symbolic moments occurred frequently along the route. When the *Graf* neared Moscow, however, and with the near-escape of the previous year fresh in his mind, poor weather led Eckener to abort a planned flyover of the city. The Russian diplomats aboard protested vigorously, threatening a diplomatic incident. Shortly after, Hearst's office received a complaint from the personal staff of none other than Joseph Stalin.

The landmarks were personal as well as political. Flying over Siberia, Eckener took the *Graf Zeppelin*, at a well-managed rate of climb, to 6000ft, the altitude that had proved fatal for the *Shenandoah*. Upon the *Graf's* arrival in Tokyo, the Japanese press, keen not to be outdone by western interlopers such as Hearst, were fulsome in their reporting. Some put the crowds that turned out to greet the ship at as many as a quarter of a million people; Emperor Hirohito was among them.

The third leg of the flight was notable, too, not least for Eckener's growing sense of showmanship. Having navigated across the Pacific Ocean, the first time an airship had done so and covering a distance of 9653km

(5998 miles), Eckener planned his arrival in San Francisco for maximum impact, floating over the Golden Gate Bridge as the sun set behind the *Graf's* great canopy, eventually landing in Los Angeles. An inversion layer of warm air made for a dicey take-off for the final leg, but thereafter the flight was uneventful. Before it returned finally to Lakehurst, the *Graf* staged a crowd-pleasing pass over Manhattan.

The success of the *Graf Zeppelin* led to increased public demand, and for still bigger behemoths. Enter, in the early 1930s, a powerful, younger sibling and stablemate, the LZ-129, better known as the *Hindenburg*. Designed for round-the-world voyages, it was, at 245m (804ft) long, only marginally larger than the *Graf*, and similarly imposing. But its four 16-cylinder Daimler-Benz engines installed it as pride of the fleet. (The *Graf* was not entirely put in the shade, and paintings of its pioneering world flight were, thanks to the influence of designer de Groot, a central feature of the *Hindenburg's* interior.)

On the face of it, the US-German Zeppelin love-in was still intact. The *Hindenburg*, another strident testament to German engineering, had been designed with one of the architectural wonders of metropolitan America very much in mind – a mooring station was located at the top of the Empire State Building. But by the early 1930s, there was more than balloons looming in the German air.

## SWASTIKAS OVER CHICAGO

With its pioneering voyage over, the *Graf Zeppelin* departed for Friedrichshafen and did not return to America until 1933. In May of that year, Chicago hosted the World's Fair. In Europe, since January, Germany had been under a new political regime, Adolf Hitler's Nazi Party.

By the early '30s, the *Graf Zeppelin* was offering passenger services to South America; it was during a return flight to Germany that it made its Chicago stopover, where it flew above the World's Fair like an old friend. Only now the friend looked quite different. In the intervening years, Germany's Zeppelins had been pressed into military service. Under orders from Hermann Goering, the Reich Minister for Aviation, the *Graf* now carried the German tricolour on its starboard side – and the Nazi swastika on its port. The German Eckener, who had built his celebrity career in America, found himself in a bind. His showman's instincts had not deserted him, however. Called upon to fly several circuits of the showground, he did so with the swastika pointing away from the crowd.

## THE END OF THE ZEPPELINS

Like the advertising blimps that came decades later, Hitler's fledgling regime quickly (➔ p.208)

Right: the *Graf Zeppelin's* 1928 flight over New York, commemorated in Germany's popular *Berliner Illustrirte Zeitung* magazine.

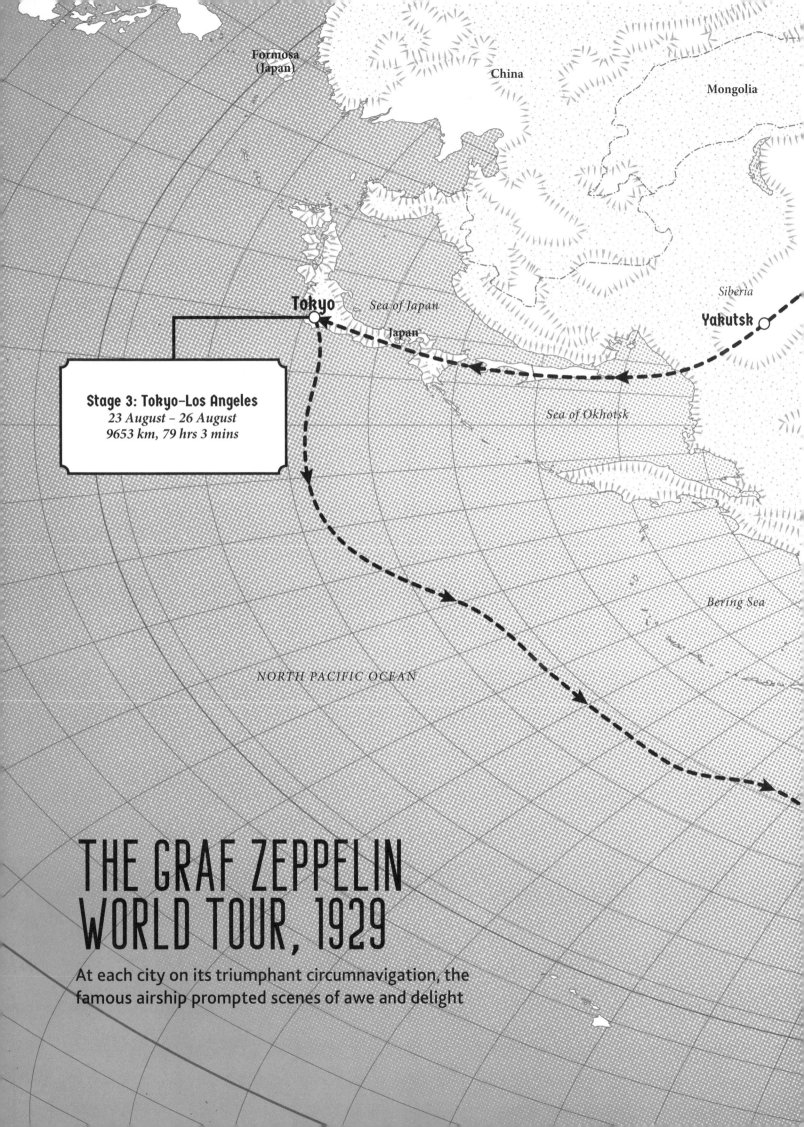

Formosa
(Japan)

China

Mongolia

*Siberia*

**Yakutsk** ○

**Tokyo** ○    *Sea of Japan*

Japan

*Sea of Okhotsk*

**Stage 3: Tokyo–Los Angeles**
*23 August – 26 August*
*9653 km, 79 hrs 3 mins*

*Bering Sea*

*NORTH PACIFIC OCEAN*

# THE GRAF ZEPPELIN
# WORLD TOUR, 1929

At each city on its triumphant circumnavigation, the
famous airship prompted scenes of awe and delight

grasped the value of conscripting Germany's Zeppelins as flying billboards. Their popularity was already established around the world, a fact not lost on Hitler's master of propaganda, Joseph Goebbels; just weeks after Hitler was installed as chancellor, on 1 May 1933, the *Graf Zeppelin* was the star turn at Germany's national Labour Day celebrations. That same month, Goebbels sent the *Graf* on a flight over Rome to mark the commencement of talks with Italy's own fascist government. Among the passengers was Italo Balbo, the Blackshirt leader who would later assume command of Benito Mussolini's air force. In September, at the Nazi Party congress in Nuremberg, the *Graf Zeppelin* assumed the role of support act, drifting above the crowds, who fell silent at precisely the moment Hitler took the stage.

Ultimately it would be the laws of physics, with a little help from market forces, that brought the reign of the Zeppelins to an end. The technology that made Germany's Zeppelins so desirable, so light, quiet and serene, was the same technology that made them not merely vulnerable but inherently dangerous.

Until 1937, the elevatorman's trade-off – maintaining gas at the ambient pressure – was considered a manageable risk. And while the *Hindenburg* employed similar release valves to the *Shenandoah*, its operating system contained one crucial difference. Whereas US balloons used helium, the optimum fuel, by the 1930s the entire German Zeppelin fleet had switched to hydrogen; lighter than air but highly flammable. In the US, helium had been in short supply for almost a decade, leading to tight restrictions on exports. (Even in 1924, there was so little helium that the national aviation authority was unable to inflate both the *Shenandoah* and a sister ship at the same time.)

## DISASTER – AND COVER-UP

On 6 May 1937, the Hindenburg was docking at Lakehurst, in New Jersey, following a routine transatlantic crossing. Without warning, and with the craft still 60m (200ft) from the ground, a portion of the great canopy exploded. With fuel reserves running to 198,200 cubic metres (7 million cubic feet) of hydrogen, the craft was enveloped in flames and fell to the ground. One of the more remarkable statistics of that tragic day is that, of the 97 people aboard, 62 managed to escape as they fell to earth, clinging to the remains of a bomb.

In a slick corporate cover-up, the fire was hastily attributed to an electrical discharge in the atmosphere. Others suggested it was an act of sabotage, scuttling the great symbol of Nazi prestige. If the survivors were one curiosity

of the *Hindenburg* disaster, another was how little the accident did in the short term to quell Zeppelin fever. Following the disaster, requests for passage to South America aboard the *Graf Zeppelin* only increased.

The German government, however, did not share the public's appetite. On the direct orders of Goering, by now formulating ideas for a resurgent Luftwaffe, plans for a *Hindenburg II* were shelved. The *Graf Zeppelin* made its final flight in late 1937, a postscript to the Zeppelin story in which it had played such a prominent part. In total, it had made close to 600 flights (a modern 747 airliner averages around 18,000 eight-hour flights in its lifetime), transporting passengers in their thousands, as well as vast quantities of mail and freight.

## RETURN OF THE DIRIGIBLE?

By the end of WWII, the Zeppelin silhouette had, once again, taken on a menacing aspect. For Londoners who survived the Blitz, or those residents of Dresden who endured extreme Allied firepower, Zeppelins brought to mind not so much a golden age of travel as the anti-aircraft blimps that had swung uselessly about their tethers, offering scant protection from high-altitude bombers. The Luftschiffbau Zeppelin, the old Count's company, was wound up in 1945.

Yet the age of the Zeppelins looks set for a return. In recent years, a number of start-ups, most notably the Aeroscraft company in California, have been raising funds and building prototypes for a new breed of high-tech dirigibles. With applications for both military and civilian use, the great gas behemoths could soon be in the skies near you.

Clockwise, from top left: the Hindenburg disaster of 6 May 1937, didn't deter passenger demand; tea break with a view for a *Graf Zeppelin* crew member; the *Graf* lifts off from Tokyo in 1929; leisure time for the paying passengers.

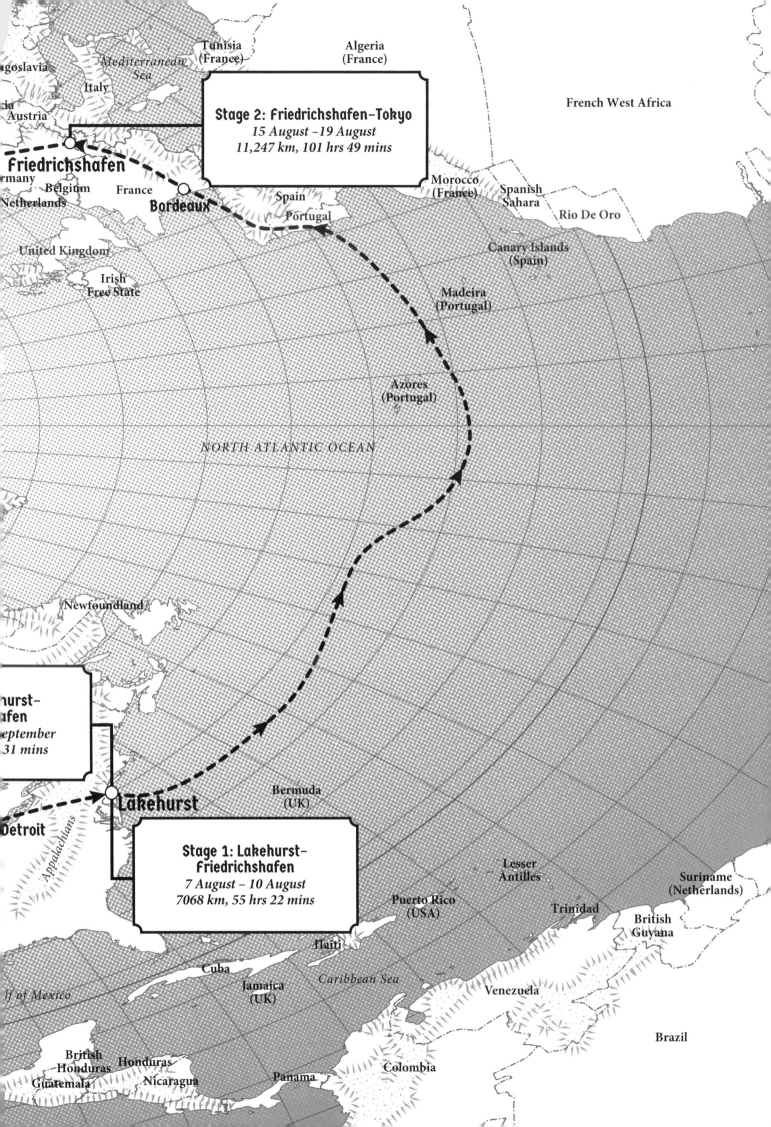

**Stage 2: Friedrichshafen-Tokyo**
*15 August – 19 August*
*11,247 km, 101 hrs 49 mins*

**Stage 1: Lakehurst-Friedrichshafen**
*7 August – 10 August*
*7068 km, 55 hrs 22 mins*

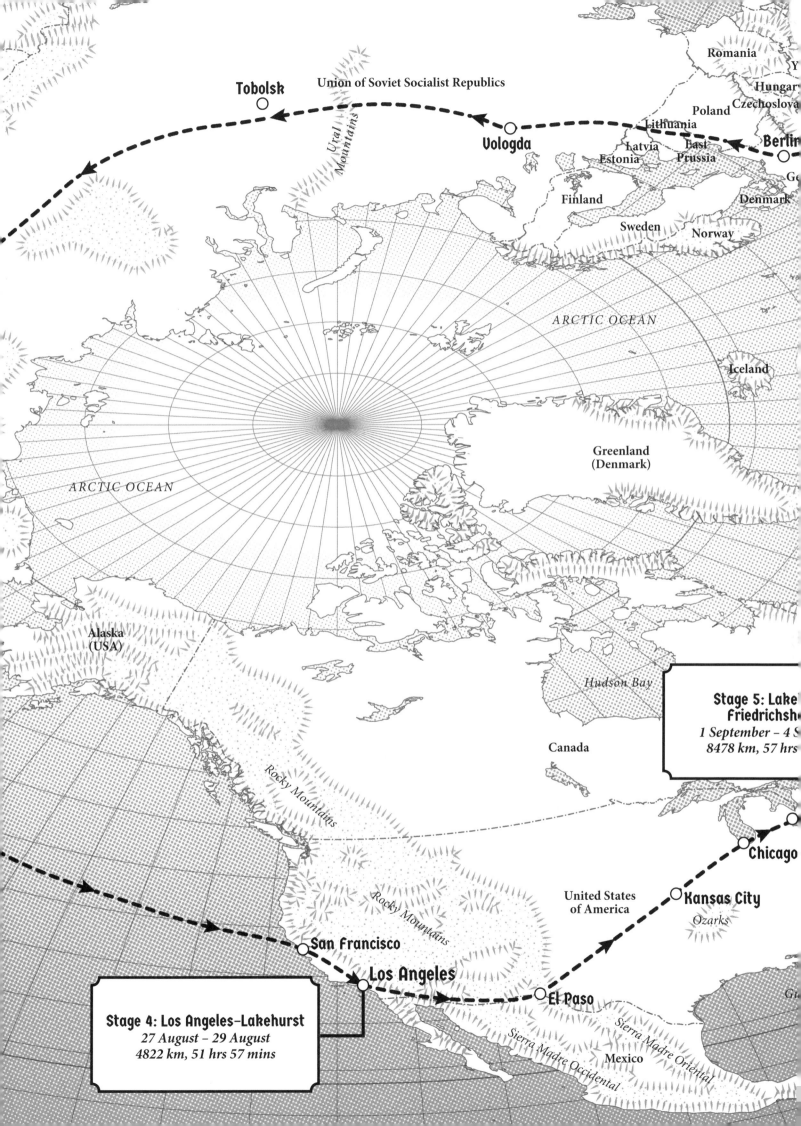

Tobolsk

Union of Soviet Socialist Republics

Vologda

Berlin

Romania

Hungary

Czechoslovakia

Poland

Lithuania

Latvia

Estonia

East Prussia

Finland

Denmark

Sweden

Norway

*Ural Mountains*

ARCTIC OCEAN

Iceland

Greenland
(Denmark)

ARCTIC OCEAN

Alaska
(USA)

*Hudson Bay*

**Stage 5: Lake
Friedrichsh**
*1 September – 4 S
8478 km, 57 hrs*

Canada

*Rocky Mountains*

Chicago

*Rocky Mountains*

United States
of America

Kansas City

*Ozarks*

San Francisco

El Paso

Los Angeles

*Sierra Madre Oriental*

**Stage 4: Los Angeles–Lakehurst**
*27 August – 29 August*
*4822 km, 51 hrs 57 mins*

*Sierra Madre Occidental*

Mexico

# UP AND AWAY: THE BEST BALLOON RIDES ON (AND ABOVE) EARTH

You don't need to soar around the world at record speed to enjoy the pleasures of ballooning. Here we offer a continent-by-continent guide to some of the planet's classic flights.

© JUSTIN FOULKES / LONELY PLANET

## EUROPE

### CAPPADOCIA, TURKEY

Cappadocia is probably the most well-known ballooning destination in the world. The region's rugged limestone peaks feature the famous 'fairy chimneys', shaped by volcanic activity, although views change with the seasons. Winter sees a carpet of snow (it's chilly hanging in the winter air, so fewer crowds, too), while spring greets an abundance of flowers, including poppies. Other architectural interests include Bronze Age dwellings and the cave churches of the Ihlara Valley.

### METEORA, GREECE

Perched upon igneous rock formations curiously untypical of the area, the monasteries of Meteora are one of the largest complexes built by the Greek Orthodox Church. Second in importance only to Mount Athos, this cliff-edge location has been occupied by Orthodox monks since the 14th century. A Unesco World Heritage Site, the region's name (an approximation of 'meteor') is derived from its lofty position. Indeed, only from the sky can you fully appreciate the scale of the monks' architectural ambition.

### LAKE GENEVA, SWITZERLAND

Lake Geneva from the air is truly spectacular. The 21,755-sq-km (8400-sq-mile) lake straddles the border of France and Switzerland and few cities on Earth can rival its Alpine location. Geneva's old town is located above the confluence of the Rhône and Arve rivers. On a clear day, the Grand Combin massif hoves into view, as does the summit of Mont Blanc. Below, the wine terraces of Lavaux, which also enjoy Unesco World Heritage status, await returning aeronauts.

Above: the Greek Orthodox monasteries of Meteora. Opposite: preparing for take off in Chile's Atacama Desert; an aerial view of the Sun and Moon pyramids at Mexico's Teotihuacán.

## 03

# LATIN AMERICA

### PATAGONIA, ARGENTINA

You'll need several thousand dollars if you want to fly above the jaw-droppingly beautiful Perito Moreno Glacier. Located in Los Glaciares National Park, in Santa Cruz, ballooning here is expensive because the region is so popular with tourists. Most visitors drive, and hike, so soaring above them is the perfect way to enjoy the 30km-long (19-mile) glacier that terminates at Argentino Lake, whose azure-tinted waters are a thing of wonder.

### TEOTIHUACÁN, MEXICO

The archaeological treasures of Teotihuacán, an ancient, Meso-American city, lie 40km (25 miles) northeast of modern Mexico City. Its sprawling nature means that only from an aerial perspective can you fully appreciate its true size. Highlights include the 3-mile-long Avenue of the Dead, which connects the site's two major pyramids, the 60m-plus (200ft) Pyramid of the Sun and the 44m (144ft) Pyramid of the Moon.

### ATACAMA DESERT, CHILE

Without significant rain for 400 years, the driest place on Earth has a stark, other-worldly landscape. Fancy life on Mars? A flight over the Atacama is as close as you'll get. In fact, Atacama's Piedras Rojas ('red plateau') has a terrain that so mimics conditions on the red planet that Nasa uses the region to test planetary rovers. The Valle de la Luna, meanwhile, is home to delicate 'wind sculptures', the astonishing result of centuries of erosion.

## 02

# NORTH AMERICA

### ALBUQUERQUE, USA

Every October, the skies above New Mexico witness one of the world's great aeronautical gatherings, the Albuquerque International Balloon Fiesta. The Fiesta draws as many as 700 balloons and 750,000 spectators from all over the world, attracted by near-perfect weather. Average annual rainfall of around 35.3cm (13.9in) makes New Mexico one of the driest states in the US. Invariably calm conditions mean the vast airborne squadrons will, at times, appear almost perfectly still.

### OTTAWA, CANADA

The Canadian capital is among a number of cities best viewed from the air. A good many flights trace the Rideau Canal, the 200km-long (124-mile) waterway that connects the city to Lake Ontario, to the south. The highlight of an Ottawa fly-by is undoubtedly Parliament Hill, known simply as 'the Hill', the complex which is home to the Canadian government. Built in 1866 in a High Victorian Gothic style, most of the original buildings were destroyed by fire, in 1916. It was rebuilt over the following decade.

# AFRICA

## NAMIB DESERT, NAMIBIA

As a ballooning destination, Namibia isn't hard to sell. Take the Sossusvlei, the sea of orange-and-pink dunes whose sharp-edged ridges dramatically announce some of the tallest sand dunes on Earth. Or the Naukluft National Park, located in Namibia's central coastal belt. At 49,768 sq km (19,215 sq miles), this is the largest game park in Africa; in addition to the gemsbok and other wildlife, it's also home to the Naukluft Mountains, a stunning massif whose fauna includes the endangered mountain zebra.

## MASAI MARA/SERENGETI, KENYA/TANZANIA

At 1510 sq km (580 sq miles), Kenya's Masai Mara is famed for its great migration, the one-million zebra, gazelle and wildebeest that pass through the park every year. A hot-air balloon offers a vulture's-eye view of the herds and prime predators they attract. The 30,000-sq-km (11,590-sq-mile) Serengeti National Park, in neighbouring Tanzania, offers similar attractions on a larger canvas. Rivers crossings are particularly tempting; in June at Serengeti's Grumeti River, and in September at the Mara.

## NILE VALLEY, EGYPT

There aren't many ways to escape the crowds at one of the world's most popular tourist destinations. But hopping into a balloon basket, preferably before sunrise, offers the chance to experience the wonders of the ancient world in unimagined tranquillity. Drift over the Pyramids, or else the famous temples of Luxor, such as Karnak or Abu Simbel. Or enjoy views of an Egypt few get to see, where farmers tend their fields and local children swim in the great river.

Above: drift over shifting, candy-coloured dunes of the Namib Desert. Right: escape the ever-present crowds with a gentle float over the Nile Valley.

# AUSTRALIA & NEW ZEALAND

## RED CENTRE, AUSTRALIA

Balloon flights are a popular way to experience Australia's ancient desert landscape. Most operators will offer dawn and/or dusk flights, the perfect vantage point from which to contemplate millennia-old rock formations, observe kangaroos and see just how close the wedge-tailed eagle can get to your canopy. As the sun rises, it lights up the MacDonnell Ranges, the mountains often depicted by the celebrated Aboriginal Australian artist Albert Namatjira.

## MELBOURNE, AUSTRALIA

The Victorian capital is the perfect ballooning location for sports fans. Landmarks to look out for from the gondola include the Melbourne Cricket Ground, and Melbourne and Olympic parks, the locations, respectively, for the annual Australian Open tennis tournament and the 1956 Olympics. And if you take a dawn flight, you can watch the city slowly come to life, as rowers undertake their morning training on the Yarra River far below. You'll also see the Grand Prix circuit snaking around Albert Park Lake.

## CANTERBURY PLAINS, NEW ZEALAND

Running from the Southern Alps to the Pacific Ocean, the fertile floodplains of New Zealand's South Island are an important agricultural area. The area accounts for around 80% of the country's arable produce, and is also notable for herbs and cut flowers. To enjoy this picture-postcard landscape, sail high above it, watching as the colours change with the light; fly in late summer or autumn, and you'll see farmers bringing in the harvest.

Below: Bagan, Myanmar's 'temple city', is a ballooning hotspot. Below right: flying low at sunrise over fields in Jaipur, India.

## 06

# ASIA

### JAIPUR, INDIA

Fly over Jaipur's famous landmarks, such as the Palace of the Winds and City Palace, and it's not hard to see why they call it the 'Pink City'. The colour denotes hospitality in India, and legend has it the buildings, on the orders of Maharaja Ram Singh, were painted pink to welcome Prince Albert, Queen Victoria's husband, in 1876. From the air, you'll also notice that most of the city isn't actually pink at all, from the outlying farms to the 16th-century Amber Fort. But Jaipur is no less spectacular for that, as an aerial tour confirms.

### BAGAN, MYANMAR

Located on a floodplain of the Irrawaddy River, the ancient city of Bagan is a ballooning hotspot. Constructed between the 11th and 13th centuries, the collapsed monasteries, stupas and pagodas are among 2000 Buddhist structures. The dry forest in which they now lie is best viewed at dawn or dusk. It might feel a bit touristy at times, although the sight of countless other balloons taking to the sky makes for quite a spectacle.

### YANGSHUO, CHINA

Yangshuo is known for its breathtaking karst scenery, particularly along the Li River, whose distinctive, conical hills are clad with temperate forests. Located in Guangxi province, Yangshuo town presents pockets of well-preserved local culture, with balloon rides gliding over rice paddies worked by buffalo. Periodically there are safety issues around some of the operators, so check before you book. Nervous flyers can also choose to navigate the Li by bamboo raft.

© ROBERT F COOKE / SHUTTERSTOCK

© TASSAPHON VONGKITTIPONG / GETTY IMAGES

## 07

# ANTARCTICA

If you're keen to soar over every continent, Antarctica is where your luck runs out. The first successful balloon flight at the South Pole was only made in 2000. A more recent attempt, to launch tourist balloons from Patriot Hills, in 2014, was cut short by heavy winds. The wider region, however, has history when it comes to inflatable aviation. In February 1902, seeking safe harbour for his ship in McMurdo Sound, Robert Falcon Scott used a balloon to ascend to 180ft. And saw nothing but ice.

# BY
# TRAIN

The romance of train travel is undeniable, but not always easy to quantify. With the exception of a few superpowered high-speed trains, riding the rails takes time, at least compared to air travel. A journey that barely gives you the chance to finish a newspaper by air can take time enough to polish off *War and Peace* by train. You can only go where the rails go, at a time dictated by connections, staffing levels and whether the weather is too hot in summer, too cold in winter or too leafy in autumn.

On the other hand, a train is a picture window, offering a journey-long sequence of framed views over unspoilt landscapes, scenes that are reduced to a topographical map from the air, or hidden behind motorway embankments and traffic when travelling by road. Compare the view when crossing the Forth Bridge in Scotland to the vista of grey traffic barriers when crossing by the adjacent road bridge.

And trains offer an unparalleled sense of momentum, of the mechanical process of getting there, that is unmatched by any other form of transport. The clatter of the wheels on the tracks measures every metre of the journey, driving you inexorably forward towards your destination. Indeed, on many rail journeys – the toy train to Darjeeling, say, or the Trans-Siberian Railway – it is the train ride that is the destination. You might be compelled to take the long way round, but when circumnavigating the globe, taking the scenic route is rather the point.

But maybe it's something less poetic. What trains offer over and above all other forms of transport is comfort. Compare a carriage on the Venice Simplon-Orient-Express to even the world's most luxurious first-class aeroplane bed-seat. And the very cheapest third-class train ticket in Asia offers the bonus of being able to get up and walk around, and even hop off the train to buy snacks on the platform – try doing that on rough seas on the crossing from Singapore to Chennai.

Little wonder then that explorers from Nellie Bly (who circled the globe in 72 days by steam train and steamship from 1889 to 1890) to Paul Theroux (whose book *The Great Railway Bazaar* inspired a million rail journeys across Asia) have expressly chosen the scenic route, putting in the meditative, exhilarating miles by rail rather than reducing the journey to a mere inconvenience by road, air or sea.

# GAUGE CHANGERS: MAKING TRACKS AROUND THE WORLD

Whether it was to wage war, tighten the grip of colonial control or accelerate the flow of riches and people, the 'railway mania' that swept the globe transformed everything in its wake: borders, empires, even time itself.

## AFRICA

There's an old Kenyan legend, attributed to the Masai and Kikuyu tribes, that the coming of a huge and monstrous iron snake would change everything in the world. For the tribes of central Kenya, this certainly proved true: when the British-built Uganda Railway opened in 1901, it precipitated the complete takeover of East Africa by colonial powers. Nevertheless, the web of railway lines that were laid across Africa during the European colonial period brought positives as well as negatives – medicine, trade, industry, employment, and of course travel, for local people as well as globetrotting sightseers.

In fact, the Uganda Railway was a relatively late arrival. The first railway line in Africa linked Alexandria and Kafr El-Zayat in Egypt in 1854, while South Africa got its first rail line in 1860. However, whether commissioned by colonial overlords or Ottoman emperors, all of Africa's early railways had one thing in common: European engineering. Experts from Britain, France, Germany and the Netherlands plotted the path of Africa's railways, while the back-breaking heavy lifting was done by crews of local people or indentured workers, shipped in from other colonial territories.

The motivation, inevitably, was financial. Despite noble official explanations, Africa's rail lines were built to transport goods to the coast, where fleets of colonial ships waited to offshore Africa's wealth for the greater glory of empire. Just as problematically, despite being built along whichever route offered the lowest cost of construction, Africa's railways came to define national boundaries, dividing tribes from their lands and communities from their coastlines. In the modern age, rail travel has expanded dramatically across the continent,

Above: the Uganda Railway stops at Maji ya Chumvi, Kenya, in 1907.

with Africa's first high-speed trains inaugurated in Morocco in 2018, but the network still maps the passage of European influence from the coast to the heart of Africa.

## ANTARCTICA

Unlikely as it might sound, Antarctica had one brief experiment with rail transportation in the early years of polar exploration. A narrow-gauge rail line was optimistically constructed to bring in supplies from the jetty at the French-run Dumont d'Urville Station, but it was quickly abandoned, being completely impractical for the frozen polar conditions.

© ALINARI / GETTY IMAGES

© ROYAL GEOGRAPHICAL SOCIETY / GETTY IMAGES

© GETTY IMAGES

## ASIA

When people talk about trains in Asia, they usually mean one thing. Nowhere caught the rail bug quite like India, and today more than 12,000 trains run daily around the country, carrying 23 million passengers, from narrow-gauge steam trains in the Himalayan foothills to the lavish, fit-for-a-maharaja carriages of the Palace on Wheels in Rajasthan.

State-owned Indian Railways has grown to become the world's eighth-largest employer, with a staggering 1.4 million staff, but the network was created to further the mercenary ambitions of the British Empire. British Governor-

Clockwise from top left: the palatial Railway Office in Calcutta (Kolkata), India, 1890; a double loop on India's Darjeeling Himalayan Railway; completed in 1899, the Ugandan Railway's Lunatic Line between Nairobi and Mombasa remained in service until 2017.

General of India Lord Hardinge gave the game away when he argued in 1843 that the railways were a vital tool for the 'commerce, government and military control of the country.'

Indian passenger services began in 1853 – not long after the first locos rumbled into France and Spain – with a 34km (21-mile) journey from Bombay to Thane, pulled by three locomotives called *Sindh*, *Sahib* and *Sultan*. Before long, rail lines extended to every corner of the country, with pint-sized trains transporting the entire colonial government to the cool foothills of the Himalayas every summer.

The rest of Asia wasn't far behind, egged on, as usual, by colonial interests. The Dutch built the first railway in Indonesia in 1867, while British-backed lines appeared in Japan (1872), China (1876) and Malaysia (1885). Today, Japan and China lead the world thanks to high-tech innovations such as Japan's *shinkansen* bullet trains, China's space-age maglev, and the world's highest rail line from Xining to Lhasa in Tibet, China.

## AUSTRALIA

The first lines in Australia were laid down before federation in 1901, linking the British-founded colonies on the coast well before the interior of the island was fully explored. Unfortunately, the rail engineers of New South Wales, Victoria and South Australia used three different track gauges, necessitating changes of train at each state border when the three networks were finally connected in the 1880s.

Getting all of Australia's railways onto a single gauge took until 1995, and the last tracks joining Alice Springs and Darwin were only laid in 2004. Apart from a handful of branch lines to mining outposts in the interior, rail expansion never strayed far from the coast. Today, just two trains cross the continent – the Indian Pacific, linking Sydney, Adelaide and Perth, and the Ghan, between Adelaide, Alice Springs and Darwin.

**CLASS ACT**
In Britain, the 1844 Regulations of Railways Act transformed rail travel into transport for the masses. Third-class passengers, at last, benefited from a roofed carriage. At least one third-class service was to run on every line, and the maximum third-class fare was a penny a mile. By 1850, third-class ticket sales accounted for half of all rail profits.

## EUROPE

Contrary to popular belief, the first train in the world was not George Stephenson's *Rocket*, but the cog-driven steam engine designed by Richard Trevithick, which completed the first ever powered rail journey at the Penydarren Ironworks in Merthyr Tydfil in South Wales in 1804. Nevertheless, it was Stephenson who brought train travel to the masses, setting himself up as the world's first commercial manufacturer of steam locomotives. The maiden passenger journey – an uncomfortable rattle on board a springless carriage known as 'The Experiment' – took place from Darlington to Stockton-on-Tees in northeast England in 1825, and was pulled by Stephenson's *Locomotion No 1*.

Where Britain led, Europe quickly followed. With the benefit of imported British know-how, railways thundered out across the continent, first in Belgium and Germany, then in Austria, Italy and the Netherlands. By 1900, the only region of Europe that did not have a rail network was Rumelia, an Ottoman province in the Balkans. The mountainous terrain of the Pyrenees and the Alps provided numerous opportunities for engineers to show off their genius with assorted rail innovations – from funicular trains to rack railways, tunnels and soaring viaducts.

The impact of the trains on European society was revolutionary. Passenger railways meant schedules and timetables, so every station needed to follow the same time. Nationwide 'standard time' was rolled out in Britain in 1840, with the US and mainland Europe quickly following suit. The railways also facilitated mass tourism as working people decamped from industrial towns to the coast to enjoy their new-found leisure time.

Across the Channel, international rail travel became the new favourite hobby of the upper classes, with plush trains such as the Orient Express whisking passengers across the continent to the 'exotic East' in unimaginable luxury. Even two world wars failed to stop the dramatic growth of the railways – if anything, demand increased as displaced people fled and troops advanced across Europe.

It's amazing what a few decades of peace can do. Modern Europe is a place of seamless rail integration, where trains thunder across international borders at speeds of up to 350km/h (215mph), with national rail companies competing to offer the new 'fastest train in Europe'. The first direct train from London to Paris took 11 hours, with the whole train loaded onto a ship to cross the English Channel; today the trip from St Pancras to Gare du Nord via the Eurotunnel takes under three hours.

Above: rail touring from London to Scotland in style, 1876.

# BY TRAIN

## THE USA

A country the size of a continent takes some conquering, and the rapid colonisation of North America was only made possible by the growth of the railways. The first rail line in the US was a British-built gravity railroad hauling freight over the Niagara Portage, but proper rail travel arrived in the 1830s with the Baltimore and Ohio Railroad, joining New York and Baltimore to the Great Lakes and Midwest.

As with other fledgling railways, the US's early railroads were founded on British tech, but the first American-built steam train, Peter Cooper's *Tom Thumb*, rolled on to the tracks in 1830. Building a railroad soon became a badge of honour for any self-respecting governor, and 'laying the first stone' and 'driving the last spike' ceremonies exploded across the country as the railways carved inexorably west.

Trains facilitated trade, moved settlers, transported soldiers, and even, in the case of the American Civil War, provided strategic advantage for the North. The east and west coasts were finally joined with the ceremonial driving of the 'golden spike', marking the opening of the First Transcontinental Railroad in 1869.

The arrival of aviation and the automobile sparked the end for rail expansion. Today, the railways limp on in the face of falling passenger numbers and shrinking funding; completion of America's only high-speed train line, the beleaguered California High-Speed Rail, has been pushed back to 2033.

## SOUTH AMERICA

With railways criss-crossing its northern neighbour, it was inevitable that South America would join the railway club. Geography provided an added incentive: ships serving the continent's biggest port at Valparaiso in Chile had to brave the deadly passage around Cape Horn to reach markets in Europe.

In 1842, steam-shipping magnate William Wheelwright was commissioned to build South America's first commercial railroad but, with the Andes in the way, it took 68 years to finally link Valparaiso to Buenos Aires. Along the way, mine-owners in the Atacama Desert became millionaires, guano-harvesters built empires in Peru, and investors in the City of London ended up controlling most of the railways in South America.

Thanks to uncooperative governments and incompatible rail infrastructure, creating an integrated rail network in South America is still an unfulfilled ambition. To this day, there are no train services between Venezuela, Colombia, Ecuador and Peru, and trains run on separate gauges in Argentina, Brazil and Uruguay.

**Below: WH Whiton's US military locomotive, 1864.**

© THE LIFE PICTURE COLLECTION / GETTY IMAGES

219

# GREAT JOURNEYS

**Across America on the Zephyr, to the steppes and beyond on the Trans-Mongolian Express, and into Australia's vast Outback aboard the Indian Pacific – on these epic services, the romance of rail's Golden Age is alive and well.**

### ACROSS AMERICA ON THE CALIFORNIA ZEPHYR

To assist Nellie Bly on her famous 1889-90 circumnavigation (see p.230), the *New York World* chartered a train from San Francisco to New Jersey. Today you can buy a standard ticket on the *California Zephyr* and get most of the way from west to east coast in three days. In fact, the *Zephyr* isn't a true coast-to-coast service; it starts inland from San Francisco at Emeryville and terminates on the shores of Lake Michigan in Chicago, but connecting trains fan out to cities along the eastern seaboard.

Starting with a short bus ride from San Francisco to Emeryville, this 3924km (2438-mile) journey takes in a broad sweep of American scenery: forests and deserts, mountains and canyons, punctuated here and there by gardens of skyscrapers. Slicing across California, Nevada, Utah, Nebraska, Iowa and Illinois, this is as close to a transect of the American West as you could hope for: the pioneer route in reverse, from San Francisco through the dramatic terrain of the Sierra Nevada and the Rocky Mountains to the Midwest.

Departing daily, the *Zephyr* takes nearly 54 hours to reach the Windy City. To help you while away the hours, there's an observation lounge with wrap-around windows – handy for the stretches where the Colorado River canyon curves overhead – plus a dining car and a choice of sleeping cars or reclining seats, depending on what your budget can stretch to.

### ESSENTIAL STOPS

The *Zephyr* runs to Chicago with only fleeting stops along the way, but travellers with time to spare can break the journey at some landmark locations. At 11.09am, the train rolls into Sacramento, with its farmers markets and

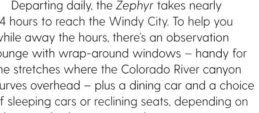

### DONNER PASS

In April 1846 George Donner left Illinois with a party of 87, for Alta California. An ill-fated deviation and October snow stranded them for the winter – when rescuers arrived six months later, one reported children 'sitting upon a log ... devouring the half-roasted liver and heart of the father'; 41 had perished, and amid accounts of treachery and cannibalism, so did the myth of the noble pioneer.

© JAN ABADSCHIEFF / 500PX

© ISTOCKPHOTO / GETTY IMAGES

pioneer attitude; by 2.38pm, you're in Truckee, gateway to Lake Tahoe.

Next, it's over the high plains to Reno, Las Vegas' casino-crammed little cousin, before an overnight push to Mormon-stronghold Salt Lake City. (Further east, at Green River, Utah, look out for an unusual sight: *Ratio*, a monumental piece of public art created from large blocks.) The scenery gets truly epic as you plunge into the canyons of the Fraser and Upper Colorado Rivers. By lunchtime, you'll hit Glenwood Springs, leaping-off point for Aspen, then Granby, the portal to Rocky Mountain National Park, before making a half-hour stop in mile-high Denver.

East of the Rockies, the train rolls past backwoods farmsteads and country towns in Nebraska, crossing the Mississippi River between Iowa and Illinois. Finally, the skyline of Chicago rises overhead like an open pop-up book around 2.50pm – just in time for a gut-busting deep-dish pizza or an Italian beef sandwich at Al's on West Taylor Street.

## ONWARD TRAVEL

Heading on to the East Coast, you'll have the afternoon and evening to spend in Chicago – easily passed at the Art Institute or taking a selfie next to Anish Kapoor's giant stainless-steel sculpture *Cloud Gate*, aka the Bean – before boarding the 9.30pm Lake Shore Limited bound for New York. It's an overnight train, so you'll miss most of the scenery between Chicago and Cleveland but catch views of Lake Erie and the Hudson River before you trundle into Penn Station, just a few miles from Hoboken Terminal (where Nellie Bly embarked on her record-breaking journey).

Continue to London by plane and you can be back on the rails by morning, bound for Paris, Moscow and Beijing.

## ALL ABOARD

You can book tickets online for the *California Zephyr* and Lake Shore Limited at the Amtrak site: www.amtrak.com.

Above: the *California Zephyr* offers a scenic journey through the US interior. Opposite, from top: the *Zephyr* whips through the epic Colorado countryside on its route between San Francisco and Chicago; rooftop windows reveal views from all angles.

© AMTRAK

221

## EUROPE TO ASIA ON THE TRANS-MONGOLIAN EXPRESS

It's possible to connect from Paris all the way to Beijing on just two trains, the Paris–Moscow Express and the Trans-Mongolian Express, creating quite possibly the world's most epic railway journey.

With a nod to Bly's original route, travel from London to Paris by Eurostar and board the weekly Paris–Moscow Express at Gare de l'Est. For the next two days, you'll get the full Russian rail experience – conductors in big hats, biscuits, tea served in glasses with *podstakannik* (metal tea-glass holders) – as you barrel across France, Germany, Poland and Belarus.

After two days and nights, you'll have earned a few days in Moscow before climbing aboard the Trans-Mongolian Express at Yaroslav Station. Seven days later, having crossed central Russia, the Gobi Desert and the stirringly empty steppes of Mongolia, you'll emerge into the reassuring cacophony of Beijing's main train station.

Along the way, you'll walk miles along train aisles, sip gallons of tea from samovars, and pass some of the most humbling scenery on the planet. The emptiness of the steppes has to be seen to be believed, though you'll spot the odd yurt to remind you that people survive in this desolate environment. There's a final daylight run through the mountains of the Yanshan Range, with glimpses of the Great Wall, before you roll into Beijing. As we said, epic.

## ESSENTIAL STOPS

Many travellers follow this spectacular route without a break; to stopover, book separate

tickets for different stretches. And what stopovers: imperial Vladimir and Nizhny Novgorod; Yekatarinburg, gateway to the Ural Mountains; Novosibirsk, Russia's third city; Irkutsk, for Lake Baikal; and Ulaanbaatar, the enigmatic capital of Mongolia. And bookending the journey, Moscow and Beijing are each worth several days of anyone's time.

## ONWARD TRAVEL

The ride doesn't have to stop at Beijing. Trains run south from Beijing to Nanning, connecting with cross-border services to Hanoi. Once in Vietnam, the Reunification Line zips south past gorgeous coastal scenery to Ho Chi Minh City. To traverse virtually train-free Cambodia, take a bus to Siem Reap via Phnom Penh, then cross the border on foot at Poipet.

Once in Thailand, normal service resumes: trains run regularly from Aranyaprathet to Bangkok, starting point for a 1920km (1190-mile) course down the peninsula to Singapore. Overnight trains rattle to the Malaysian border at Padang Besar, where the modern, electrified Malaysian rail network whooshes south to Butterworth (for Penang) and Kuala Lumpur. From KL, things get finicky. To reach Singapore, take a train to Pulau Sebang or Gemas and change for Johor Bahru, then take a bus or taxi across the causeway.

## ALL ABOARD

Book tickets through Russian Railways (www.rzd.ru) or online through an agency such as Real Russia (https://trains.realrussia. co.uk/transsib).

Opposite, from top: Gobi Desert camels; fancy fretwork in the dining car of the Trans-Mongolian Express. Above: China's Great Wall. Below: Ulaanbaatar's Genghis Khan statue, Mongolia.

## LAKE BAIKAL

Northeast of Irkutsk, southern Siberia, lies the oldest, largest (by volume) and deepest (1632m / 5354ft) lake in the world, formed when a rift valley appeared some 25 million years ago. So huge is the lake that it's said to contain 20% of the planet's lake and river water. It's calculated that a water molecule typically takes over 300 years to progress from inlet to outlet. Baikal is also a habitat for 1500 animal species, including the nerpa, the only freshwater seal.

## COAST TO COAST ON AUSTRALIA'S INDIAN PACIFIC

Touching down in Perth by plane, you have a country as large as a continent to conquer. The railway line hugging the centre of Australia like a waistband was only completed in 1970, when three incompatible gauges of track were brought onto a single system, and it remains the only truly transcontinental rail journey in the world – nowhere else will a single ticket take you from coast to coast, right the way across a continent.

Still using the original stainless-steel carriages from the 1970s, the Indian Pacific cuts a 4343km (2699-mile) slice through the Australian Outback, running every Sunday from Perth to Sydney via Adelaide (and in the opposite direction on Wednesdays). Berths are beds in comfortable but cramped roomettes, with hot showers at the end of the carriage, and there's a restaurant car for fine dining (included in the not-insignificant ticket price) and a lounge bar for whiling away the hours with chat and cocktails. Think of it as a small but luxurious hotel on wheels, with a mandatory three-night, four-day stay.

The reason travellers put up with these compact conditions is the scenery: eucalyptus forests on the slopes of the Blue Mountains, arid bush country around Broken Hill and the eerie emptiness of the Nullarbor Plain, where the highest point on the skyline is a passing kangaroo (the clue is in the name – Nullarbor means 'no trees').

En route, you'll cross the longest section of straight railway in the world, running straight as a ruler for 478km (297 miles). Eventually, after days of dust, deserts and more emus, kangaroos and cockatoos than people, you'll clickety-clack into Sydney Central Station.

## ESSENTIAL STOPS

Conveniently for travellers, the Indian Pacific makes stops that are long enough for passengers to get off the train and explore. Leaving from Perth, there's time for an evening drink in the gold-rush town of Kalgoorlie, a leg-stretch in the ghost town of Cook, a slap-up breakfast and downtown walkabout in Adelaide, and a peek at the miners' memorials in Broken Hill. The train only runs once a week

### NUCLEAR FALL-OUT

Midway along the Indian Pacific, the train passes Maralinga, one of three Australian sites where the British government tested atomic weaponry in the 1950s. Indigenous peoples and military personnel suffered incalculably, and the expensive clean-up lasted many decades.

© MATT MUNRO / LONELY PLANET

of course, so remember not to wander too far from the station!

For something considerably more ambitious, ride the route halfway to Adelaide, stopover from Tuesday to Sunday, and connect to the similarly luxurious *Ghan* service, running north through the Red Centre to Alice Springs and on up to Darwin.

## ONWARD TRAVEL

Perhaps you want to stitch together all of the rail journeys described in the previous few pages and create your own circumnavigation of the world by travelling to San Francisco. Travelling on from Sydney, your rail options are curtailed by the Pacific Ocean, however. The next nearest railways are a handful of heritage lines run for tourists in Hawaii – a worthy stopover on the flight to San Francisco to finish where you started. .

## ALL ABOARD

Book tickets on the Indian Pacific at www.journeybeyondrail.com.au/journeys/indian-pacific.

Far left: Australia's Indian Pacific Railway pulls into Bathurst station.

This page: Clockwise from top: rural New South Wales from the window of an Indian Pacific carriage; lunch is served in the Queen Adelaide restaurant car; the Indian Pacific makes a stop at Watson on the Nullarbor Plain.

225

# WHERE STEAM IS STILL HOT

**Across the planet, the oldest locomotive technology of them all continues to reach the places that diesel and electric never will.**

Y ou don't have to be a trainspotter to carry a torch for the golden age of steam. The modern world as we know it was built on the steam engine – in simple terms, a kettle with a piston – and riding a vintage steam locomotive is still one of the most evocative ways to travel. It could be the noise, the vibration, the smell, the chesty whoomph of steam bellowing out of the chimney or just the honest, mechanical process of getting from A to B.

The country to use steam trains as the backbone of its rail infrastructure until most recently was **China**. The last Chinese-made steam locomotive rolled out of the factory

in 1999, a few years after the bulk of India's enormous, charismatic steam trains were pulled from the rails and shunted off to the scrapyard. Today, just a handful of heritage lines continue to clatter through the countryside, attracting legions of enthusiasts desperate for a final taste of the glory years of rail travel.

In China, the Jiayang Coal Railway is the last commercial steam service in the country, running 20km (12 miles) from Shixi to the Jiayang Coal Mine. A living relic from the Great Leap Forward, the diminutive loco was commissioned in 1959 and it still provides crowded daily passenger services for local mineworkers, who squeeze onto hard wooden benches while a fine rain of ash blows in through the glassless windows. For less hardy travellers, a sightseeing

Left: China's pint-sized Jiayang Coal Railway. Above: the Trochita chugs through the Patagonian Andes. Opposite: Glenfinnan Viaduct in Scotland.

service that's gentler on the hindquarters runs daily between Yuejin and Huangcun Mine.

**India** has also managed to cling on to a handful of its once vast fleet of steam locos, for reasons of practicality as much as any particular passion for steam. The rail lines that climb into the Western Ghats and the foothills of the Himalayas were built using three obsolete narrow gauges – 1000mm (3ft 3⅜in), 762mm (2ft 6in) and 610mm (2ft) – and successive Indian governments have baulked at the cost of forging new standard gauge lines through this unforgiving terrain. So, charming, pint-sized trains such as the Darjeeling Himalayan Railway, the *Himalayan Queen* to Shimla and the rack-based Nilgiri Mountain

Railway to Ooty (Udhagamandalam) still chuff through the steep hills, pulled by a mix of steam and diesel engines.

India can also lay claim to the oldest working steam locomotive – the *Fairy Queen*, built as East Indian Railway Nr. 22 in 1855. This stocky, wrought-iron beauty was manufactured in Leeds, and transported by steamship to Calcutta, seeing service in West Bengal and Bihar and carrying troops during the First War of Indian Independence. The venerable loco still rolls periodically from Delhi to Alwar in Rajasthan, resting between times in the Railway Heritage Museum at Rewari in Haryana.

Many other former colonies still fly the flag for steam. **Indonesia** keeps a hand in the

© ALAMY STOCK PHOTO

game with the Cepu Forest Railway, an old logging line from 1915 which continues to trundle through the teak plantations surrounding the town of Cepu in Java. Across the Indian Ocean in **South Africa**, the lavish, steam-hauled trains of Rovos Rail connect Cape Town, Durban and Pretoria, while 22 nostalgic narrow-gauge locos shuffle around the Sandstone Estate near Ficksburg.

You don't have to travel to the ends of the Earth to find lingering steam railways. Hauled by a rotating fleet of vintage engines, the *Jacobite* in **Scotland** chugs from Fort William to Mallaig across the Glenfinnan Viaduct, made famous by cameos in the Harry Potter movies. A dozen other steam trains rumble through the green

and pleasant British countryside, including the tiny rack-and-pinion railway that climbs the rugged slopes of Mt Snowdon in **Wales**.

Similar heritage lines keep the dream of steam alive across Europe, from **Russia**'s extravagant *Golden Eagle* on the Trans-Siberian Railway to **Germany**'s Brocken Railway and **France**'s Froissy Dompierre Light Railway, which traces a humbling route through the battlefields of the Somme.

If you've ever dreamed of the Old West, seek out the tangle of heritage rail lines that cross the **USA**, where you can thunder through the countryside, imagining hot pursuit by train robbers on horseback. Hulking, black steam locos haul passengers along precarious mountain ledges on the Durango & Silverton Narrow Gauge Railroad and the Cumbres & Toltec Scenic Railroad in Colorado and New Mexico.

Further south, the Trochita – aka the Old Patagonian Express – rolls along a section of its former 402km (250-mile) route through the Andes in **Patagonia**, while over in **Brazil**, the *Maria-Fumaça* ('Smoking Mary') still trundles along the 13km (8 miles) of track between São João del Rei and Tiradentes in Minas Gerais.

Clockwise from top: the diminutive Darjeeling Himalayan Railway navigates north India's mountainous terrain; vintage locos at the Heritage Steam Engines Parade in New Delhi; the Darjeeling Himalayan at West Bengal's epic Batasia Loop.

# NEVER MIND PHILEAS FOGG, HERE COMES NELLIE BLY

When the New York World's star reporter declared she could circumnavigate the globe by steam in less than 80 days, her editor scoffed 'No one but a man can do this!'

Elizabeth Cochrane Seaman was a pioneer in more ways than one. The groundbreaking feminist, writer, traveller and investigative journalist pushed back the limits of what women were perceived to be capable of, in a country where men determined the rules, wrote the records and controlled all the boundary markers.

She was also the first person to circumnavigate the globe by train, some six decades after the world's first locomotives first took to the rails.

## FROM RAGING TO WRITING

Born to a family of mill-owners in Pittsburgh in 1864, Seaman nevertheless had to abandon teacher training through lack of funds. She first made waves as the writer of an excoriating letter about the coverage of women in the *Pittsburg Dispatch*, under the pseudonym 'Little Orphan Girl'. The editor was so impressed that he ran a campaign to find the author of the note, and promptly offered her a job.

Under the patriarchy of the time, women writers were obliged to use pen names, so Seaman chose Nellie Bly, borrowed from a popular music-hall song of the time.

## THE ONLY SANE PERSON IN THE ASYLUM

Bly moved to New York in 1887, seeking a more rewarding role in journalism – something beyond the women's pages, to which she'd been eventually relegated at the *Pittsburg Dispatch*, writing about women's issues, fashion and gardening.

Her technique for gaining attention was certainly ground-breaking: working undercover for the *New York World*, Bly feigned madness and had herself admitted to the asylum on Blackwell's Island, writing a no-holds-barred exposé of the treatment of the mentally ill on her release.

Her report, 'Ten Days in a Mad-House', was one of the first ever pieces of investigative journalism, propelling Bly to the upper ranks of American journalism. In it, she wrote of beatings and cruelty, filthy conditions, unsanitary food and the psychological torture of being forced to sit for days in freezing conditions on bare wooden benches, deprived of all knowledge of the outside world. It was an experience, she wrote, that in two months would make any woman a 'mental and physical wreck'.

## THE BIG IDEA

Bly was clearly made of tough stuff, and the following year, she told her editor at the *New York World* that she intended to circumnavigate

Opposite: journalist and adventurer Elizabeth Cochrane Seaman – better known as Nellie Bly – set a new world record when she circumnavigated the globe in just 72 days.

the globe by steamship and train, recreating the journey of Phileas Fogg in Jules Verne's novel *Around the World in Eighty Days*. Editor Joseph Pulitzer initially baulked, exclaiming 'no one but a man can do this!'. But Bly's riposte – 'Start the man, and I'll start the same day for some other newspaper and beat him' – left him with little choice but to accede.

Bly reportedly came up with the idea for a jaunt around the globe at 3am on a Monday morning, and immediately put her plan into action. 'I went to a steamship company's office that day and made a selection of timetables', she wrote. 'Anxiously I sat down and went over them and if I had found the elixir of life I should not have felt better than I did when I conceived a hope that a tour of the world might be made in even less than 80 days.'

## BEATING THE COMPETITION

While preparations were underway, unbeknownst to Bly, rival newspaper *Cosmopolitan* commissioned its own reporter, Elizabeth Bisland, to leave on the same day, travelling in the opposite direction, turning the circumnavigation into a battle not just against time but between media moguls. (Bly was oblivious of her competitor until she arrived in Hong Kong, China whereupon she claimed not to care: 'I promised my editor that I would go round the world in 75 days ... If someone else wants to do the trip in less time, that is their concern.')

A year after making her ambitious pitch, Bly boarded the steamship *Augusta Victoria* at Hoboken, New Jersey, on 14 November 1889,

bound for Southampton, England, carrying little more than the clothes she stood up in. It was the start of a journey that would take Bly over the Atlantic, across Europe, through the Middle East, over the Indian Ocean to Asia, and back across the Pacific to North America, rolling back from San Francisco into New Jersey just 72 days after she left the jetty.

Bly's packing technique should be a lesson to anyone who struggles with an overladen backpack. She left America with the dress she was wearing – specially commissioned from a dressmaker to stand 'constant wear for three months' – plus a spare dress for warmer climates, a winter coat, some changes of underwear and a small handbag of toiletries. To fund her trip, Bly carried around £200 ($260) in English banknotes and gold and some US bills – in total, the equivalent of around $32,000 (£25,000) today – hidden in a pouch worn under her clothes. She declined the suggestion made by a male colleague to carry a revolver.

## THE 'GUESSING MATCH'

International travel at the time was far from the comfortable experience enjoyed today. Steamships sank, trains derailed, travellers vanished and illness stalked the world's travel routes like a phantom. The first international phone call only took place in 1881, and contacting home was a luxury only possible by telegram in a few strategic locations. Bly was put in touch with diplomats and journalists living in cities along the route, but after leaving New York, she was pretty much on her own.

## BLY VS BISLAND

Nellie Bly's rival, Elizabeth Bisland, was, at mid-point, three days ahead of her. How did Bly triumph? It's said that Bisland missed a swift transatlantic ship home from Southampton, England, because she was told, incorrectly, that it had already departed. In fact, her publisher had bribed the shipping company to ensure Bisland boarded. She failed to, catching a slower ship home instead, losing crucial time to Bly. Misfortune? Incompetence? Or skulduggery?

Clockwise from top left: Victorian London; Jules Verne in 1900; illustration from *Around the World in Eighty Days*; Nellie Bly.

To maintain interest in the trip, the *New York World* organised a competition, the 'Nellie Bly Guessing Match', inviting readers to guess the time Bly would arrive back home to the second, with an all-expenses paid trip to Europe as the grand prize. On the first Sunday the competition ran, the newspaper received more than 100,000 entries. By the following week, circulation of the paper had increased by 300,000 readers. Bly's trip was the talk of the land.

## THE ROUTE

Circumnavigating the globe by rail at this time was a rather different experience to circling the planet by train today. For one, there were fewer rail lines. A few decades later, Bly could have travelled most of the way from Europe to Asia by train, but in 1889, the world's rail networks were disjointed, with lines stopping abruptly at national borders, and trains running on different gauges in neighbouring provinces.

Adding to the logistical complications, obstacles such as the Atlantic and Pacific needed to be conquered, and parts of the Near and Far East were closed to outsiders. Unavoidably, Bly completed many stages of her trip by steamship, connecting the crossings by rail, rickshaw, horse and donkey. Nevertheless, even relying on steam power and coal, Bly had a carbon footprint that would shame most people crossing the planet by plane today.

## A NERVY START

Despite her steely ambition, the writer was not immune to nerves. As the *Augusta Victoria* slipped away from the quayside at Hoboken, she wrote of her fears: 'Intense heat, bitter cold, terrible storms, shipwrecks, fevers, all such agreeable topics had been drummed into me until I felt much as I imagine one would feel if shut in a cave of midnight darkness and told that all sorts of horrors were waiting to gobble one up.'

The crossing to Southampton, however, was uneventful, excluding bouts of seasickness that were to become a recurring theme for the trip. For the seven slow days it took to reach England, Bly passed her time writing, talking to other passengers, dining at the captain's table and vomiting over the ship's rails.

## THE RUSH TO AMIENS

On arrival, the peripatetic correspondent was met by the *New York World*'s London writer, bearing a personal invitation to dinner with Jules Verne at his home in Amiens. Getting to the dinner on time entailed a fevered rail journey across England and France, with barely time to pause for food or rest for two days and nights.

With no time to spare, Bly connected to London the same night on a passenger

carriage, attached at short notice to a mail train, with just a cast-iron foot warmer for comfort. On arrival at Waterloo, a Hackney carriage sent by the newspaper whisked Bly through the foggy streets to the offices of the American Legation, where a passport was handwritten on the spot.

Bly's next stop was the London office of the Peninsular and Oriental Steam Navigation Company, to stock up on boat tickets through the Far East, before zipping to Charing Cross Station to board a steam train to the coast for the crossing to Boulogne in northern France

With Jules Verne's dinner invitation pressing, Bly had no time to dally

Above and below: Brindisi harbour, circa 1869 and Brindisi harbour, circa 1900.

## DINNER WITH JULES VERNE

Bly's account of her first train travel outside America was fiercely critical of both the comforts and the character of British locomotives. 'If it is privacy the English desire so much,' she wrote, 'they should adopt our American trains, for there is no privacy like that to be found in a large car filled with strangers. Everybody has, and keeps his own place. There is no sitting for hours, as is often the case in English trains, face to face and knees to knees with a stranger, offensive or otherwise, as he may chance to be.'

Another cold and uncomfortable train journey from Boulogne delivered the outspoken journalist to the home of Jules Verne, whose 1873 novel had provided the timetable for circumnavigation that Bly was hoping to beat. Speaking just a few words of French, Bly relied on a translator for a whirlwind conversation that covered everything from Verne's inspiration for writing *Around the World in 80 Days* (it turned out to be a chance article spotted in a French newspaper), as well as his route recommendations (he advised travel via Bombay, rather than Bly's chosen route via Ceylon), and his sadness at no longer being able to trot the globe due to ill health.

## AU REVOIR, EUROPE

Time waits for no man, or woman, and Bly rushed straight from the Vernes' home to the station at Amiens to board the Club Train to Calais, then one of the most luxurious trains on the continent, with a dining car and drawing room where passengers gathered to smoke over coffee after dinner. At Calais, Bly embarked on her longest rail journey outside the USA, a non-stop, 2000km (1240-mile) ride to Brindisi on the heel of Italy – the most practical port of embarkation for the steamship to North Africa, bypassing the troubled southern Balkans, then part of the imploding Ottoman Empire.

For Bly, the journey was the destination, and she wrote little about what she saw of Europe through the train windows, and much about the state of the trains and the foibles of her fellow passengers. The schedule was always at the front of her mind, with many tight connections. Arriving two hours late in Brindisi, Bly narrowly avoided missing the boat to Port Said after rushing into town to send a cable to New York.

## TO AFRICA AND BEYOND

Aboard the steamship *Victoria*, Bly did not warm to the colonial Brits bound for Raj-controlled India. But other aspects of the crossing to North Africa appealed: 'The balmy air, soft as a rose leaf, and just as sweet, air

Below left: central Colombo, Ceylon (Sri Lanka) in 1910, with the Grand Oriental Hotel on the right. Below right: Bly had become a celebrity by the end of her circumnavigation.

such as one dreams about but seldom finds', she wrote. 'Standing there alone among strange people, on strange waters, I thought how sweet life is!'

Bly weathered the unsociable company on the *Victoria* as best she could. Some days into the journey, a rumour started amongst fellow passengers that she was an eccentric American heiress, 'traveling about with a hair brush and a bank book', and she found herself beset by would-be suitors enamoured with the thought of her (non-existent) fortune.

## FIRST IMPRESSIONS OF EGYPT

Arrival in Port Said was a chaotic free-for-all, as boatmen crowded the *Victoria* while passengers and crew tried to fend them off with poles, walking canes and umbrellas. 'One of the Arabs told me that they had many years' experience in dealing with the English and their sticks,' she wrote, 'and had learned by bitter lessons that if they landed an Englishman before he paid they would receive a stinging blow for their labor.'

On shore, Bly encountered beggars, veiled women, camels and even a captured crocodile, before continuing south along the Suez Canal, recently seized from the French by the British Empire. The opening of the canal was one of the triggers for the 'Panic of 1873',

**Above:** Mount Lavinia Hotel, Ceylon (Sri Lanka), in 1890. **Left:** portrait of Nellie Bly, January 1880.

which almost wiped out the American railway industry, after British sea trade slumped due to overdependence on sailing ships that were unable to use this game-changing shortcut to the Indian Ocean and Asia beyond.

## TRAVERSING THE SUEZ CANAL

At Ismailia, halfway along the canal, Bly visited the palace of the Khedive, the viceroy of Ottoman-controlled Egypt, but mostly she wrote of the stifling heat, occasional glimpses of Bedouin nomads and the monotonous scenery of the blank sand walls of the canal. Things picked up as the ship entered the Bay of Suez; enterprising *felucca* captains climbed aboard, hawking fruit, tropical seashells and souvenir snapshots, and performing conjuring tricks for tips.

Traversing the Red Sea to Aden, passengers turned to increasingly desperate diversions such as glee singing to distract themselves from the heat. Bly went ashore again at Aden, at the time a cosmopolitan city of Arabic and African Muslims, Jews and Iranian Parsis. Like many travellers before her, she marvelled at the personal ornamentation she encountered: men with jangling finger and toe rings and hair bleached yellow with lime; women with

Left: Galle Face in Colombo, Sri Lanka, as it is today.

earlobes stretched almost to their shoulders by huge gold earrings.

## BY SEA TO COLOMBO

On the sailing onwards to Colombo, passengers attempted to impose some normality onto the journey, passing the days creating *tableaux vivants* (living 'paintings' with human models in costumes) and watching magic lantern shows, while the warm waters of the Indian Ocean slipped by. It was a slow, tedious journey; by the time she reached the Ceylonese capital, Bly couldn't wait to get ashore.

Travelling to the jetty on a traditional catamaran (a word derived from the Tamil term *kattumaram* meaning 'tied wood'), Bly made straight for the lobby of the Grand Oriental, then Colombo's finest hotel. 'In this lovely promenade the men smoked, consumed gallons of whiskey and soda and perused the newspapers,' she wrote, 'while the women read their novels or bargained with the pretty little copper-colored women who came to sell dainty hand-made lace, or with the clever, high-turbaned merchants who would snap open little velvet boxes and expose, to the admiring gaze of the charmed tourists, the most bewildering gems.'

After sampling her first curry – a shrimp masala, served with sides of Bombay duck, beaten rice and spice-dusted fruit slices – Bly spent days exploring Colombo by carriage, hand-pulled rickshaw and bullock-cart, promenading on the shore at Galle Face Green public park and in the grounds of the Mount Lavinia Hotel, watching snake charmers, visiting temples and slipping out to local theatres in the backstreets of Colombo. She travelled to Kandy by steam train, dropping into the Temple of the Sacred Tooth Relic (which didn't impress her) and wandered the green bower of the Royal Botanic Gardens at Peradeniya (which did).

## ONWARDS THROUGH THE STRAITS

In fact, her week-long stay in Colombo was enforced – the *Oriental*, the ship carrying Bly onwards to China, was delayed by the late arrival of the connecting *Nepaul* from Calcutta, setting her schedule back by days. Steaming on to Penang in the British-governed Straits Settlements, Bly had just six hours to explore the city famed as the Pearl of the Orient – time enough, though, to explore Hindu temples and Confucian joss houses and sip Chinese tea with local shopkeepers.

Cruising south to her next stop in Singapore, Bly bemoaned the sticky humidity. 'It (→ p.241)

(→ p.241)

## THE RAILWAY CRASH

The railroads were a crucial part of the US economy after the Civil War: 56,330km (35,000 miles) of new track were laid between 1866 and 1873, making rail the country's largest non-agricultural employer. But fast, expensive expansion was risky. When Jay Cooke and Company, a banking firm invested in rail, went bankrupt in 1869, there was crisis: 89 railroads went bankrupt, a further 18,000 businesses failed within two years, and by 1876 the unemployment rate was 14%.

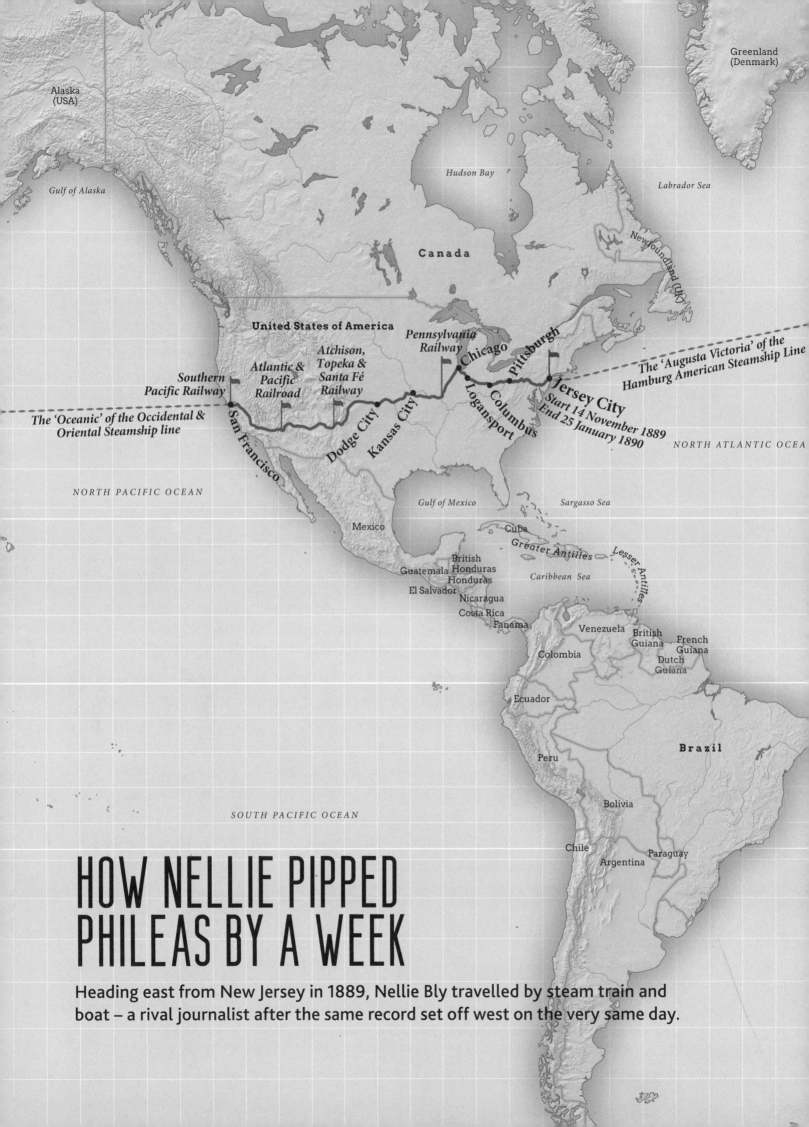

Alaska
(USA)

Gulf of Alaska

Hudson Bay

Canada

Greenland
(Denmark)

Labrador Sea

Newfoundland (UK)

United States of America

Pennsylvania
Railway

Chicago

Pittsburgh

The 'Augusta Victoria' of the
Hamburg American Steamship Line

Atchison,
Topeka &
Santa Fé
Railway

Atlantic &
Pacific
Railroad

Southern
Pacific Railway

Columbus

Logansport

Jersey City
Start 14 November 1889
End 25 January 1890

The 'Oceanic' of the Occidental &
Oriental Steamship line

San Francisco

Dodge City

Kansas City

NORTH ATLANTIC OCEAN

NORTH PACIFIC OCEAN

Gulf of Mexico

Sargasso Sea

Mexico

Cuba

Greater Antilles

Lesser Antilles

British
Honduras

Guatemala
Honduras

Caribbean Sea

El Salvador

Nicaragua

Costa Rica

Panama

Venezuela

British
Guiana

French
Guiana

Colombia

Dutch
Guiana

Ecuador

Brazil

Peru

SOUTH PACIFIC OCEAN

Bolivia

Chile

Argentina

Paraguay

# HOW NELLIE PIPPED PHILEAS BY A WEEK

Heading east from New Jersey in 1889, Nellie Bly travelled by steam train and
boat – a rival journalist after the same record set off west on the very same day.

# CALLING AT ALL STATIONS...

How quickly could you circumnavigate the world by train?

 EUROPE

 ASIA

## MOSCOW

### LONDON TO MOSCOW

**Fastest**
London to Paris by
Eurostar: 2hr 15m
Transfers by Metro: 1hr
Paris to Moscow by
Paris-Moscow Express:
1½ days

**Slowest/cheapest**
London to Brussels
by Eurostar: 2hr

Brussels to Cologne by
ICE or Thalys train: 3hr
Cologne to Berlin by
ICE train: 5hr
Overnight in Berlin
Berlin to Warsaw by
Eurocity train: 6hr
3hr stopover
Warsaw to Moscow on
the *Polonez*
sleeper: 15hr 30m

### MOSCOW TO HANOI

Moscow to Beijing via
Ulaanbaatar on the Trans-
Mongolian Express: 130hr
Beijing to Nanning on Train
G529/G421: 10 to 13hr
Nanning to Hanoi on Train
T8701: 11½hr

### SINGAPORE TO BALI

Singapore to Sekup
by ferry: 1hr
Sekupang to Tanjun
Buton by ferry: 5hr
Tanjun Buton to Pel
by minibus: 4hr
Pekanbaru to Palen
by bus: 19½hr
Palembang to Band
Lampung by train:
Bandar Lampung t
Bakauheni by bus:
Bakauheni to Mera

## LONDON

## ISTANBUL

### LONDON TO ISTANBUL

**Most luxurious**
London to Paris by
Eurostar: 2hr 15m
Transfers by Metro: 1hr
Paris to Istanbul on
the Venice Simplon-
Orient-Express: 6 days,
including a night in
Bucharest

**Slowest/cheapest**
London to Paris by
Eurostar: 2hr 15m
Transfers by Metro: 1hr
Paris to Munich
by TGV: 8hr
Munich to Budapest by
EuroNight Sleeper: 7½hr
Budapest to Bucharest
by sleeper train *Ister/*

*Dacia*: 17hr
24hr stop in Bucharest
Bucharest to Istanbul on
the sleeper *Bosfor*: 13½hr

**Alternative route**
London to Paris by
Eurostar: 2hr 15m
Transfers by Metro: 1hr
Paris to Munich by TGV: 8hr
Munich to Zagreb by
sleeper *Lisinski*: 9hr
Zagreb to Belgrade by
EuroCity train: 6½hr
Belgrade to Sofia on
the *Balkan*: 11hr
24-hour stop in Sofia
Sofia to Istanbul on the
*Sofia-Istanbul Express*:
9hr 10m

### ISTANBUL TO HANOI

Istanbul to Ankara by
YHT train: 4hr 15m
Ankara to Kars on the
Doğu Ekspresi: 24½hr
Kars to Batumi by bus:
5hr
Batumi to Tbilisi
by *Stadler Kiss*
train: 5hr
Tbilisi to Baku on Train
37: 12½hr
Baku to Aksaraiskya on
Train 55: 21hr 15m

Aksaraiskya to
Atyrau on
Train 334: 6hr
Atyrau to Almaty
on Train 29: 38hr
Almaty to Urumqi on
Train K9796: 23½hr
Urumqi to Guǎngzhōu
on Train Z232: 48hr
Guǎngzhōu to Nanning
by bullet train: 4hr
Nanning to Hanoi on
Train T8701: 11½hr

### HANOI TO

Hanoi to
City on th
*Express*: 3
Ho Chi Mi
to Siem R
bus: 13hr
Siem Rea
by bus: 3h
Walk Poip
Aranyapr
cross bord
Aranyapr
Bangkok
280/276:

## HANOI

Alaska
(USA)

*Bering Sea*

*Sea of
Okhotsk*

*Sea of
Japan*

**Qing Empire**

*Yellow
Sea*

Japan • **Yokohama**

*The 'Oceanic' of the Occidental &
Oriental Steamship line*

*East
China
Sea*

*NORTH PACIFIC OCEAN*

**Canton**

Taiwan
(Japan)

**Hong Kong**

Hawaii
(USA)

*Philippine Sea*

**Siam**

*The 'Oriental' of the
Peninsular and Oriental
Steamship Line*

*South
China
Sea*

Philippines
(USA)

*aman
ea*

*Celebes
Sea*

nang •

Malaya (UK)

**Singapore**

East Indies (Netherlands)

*Java Sea*

*Banda Sea*

New Guinea
(Germany)

Solomon
Islands
(UK)

Portuguese
Timor

*Arafura Sea*

Papua (UK)

*Timor
Sea*

New Hebrides
(France, UK)

Fiji
(UK)

*Coral Sea*

New Caledonia
(France)

**Australia (UK)**

*SOUTH PACIFIC OCEAN*

*Tasman Sea*

New
Zealand
(UK)

ng | by ferry: 2hr
Merak to Jakarta by train: 4hr
Jakarta (Pasar Senen) to Bangil by train: 12½hr
nbaru
Bangil to Banyuwangi (Karangasem) by train: 5hr 15m
pang
ar
Banyuwangi (Ketapung jetty) to Gilimanuk (on Bali) by ferry: 1hr
to 10hr
hr
Gilimanuk to Denpasar by bus: 3hr

SINGAPORE

BALI

NGAPORE
Chi Minh
Reunification
hr
h City
ap by
to Poipet
et to
het: 1hr to
er
het to
n Train
to 5hr

Bangkok to Padang Besar on Train 45: 17hr 45m
Padang Besar to Kuala Lumpur by ETS train: 5½hr
Kuala Lumpur to Gemas by ETS train: 2½hr
Gemas to Johor Bahru by Intercity or Shuttle train: 5hr
Johor Bahru to Singapore by bus: 1 to 2hr

## AUSTRALIA

### PERTH TO SYDNEY
Perth to Sydney on the Indian Pacific: 4 days, 3 nights (75hr)

### DARWIN TO SYDNEY
Darwin to Adelaide on the *Ghan*: 3 days, 2 nights
Adelaide to Melbourne on the *Overland*: 10hr 30m
Melbourne to Sydney on the XPT: 11hr

© MATT MUNRO / LONELY PLANET

## NORTH AMERICA

### SAN FRANCISCO TO NEW YORK
San Francisco to Emeryville by bus: 1hr
Emeryville to Chicago on the *California Zephyr*: 51½hr (6hr 30m stop in Chicago)
Chicago to New York on the Lake Shore Limited: 20hr

### SAN FRANCISCO TO BOSTON
San Francisco to Emeryville by bus: 1 hour
Emeryville to Chicago on the *California Zephyr*: 51½hr (6½-hour stop in Chicago)
Chicago to Boston on the Lake Shore Limited: 22hr 40m

### LOS ANGELES TO NEW YORK
Los Angeles to New Orleans on the Sunset Limited train: 45hr (9hr 20m overnight stop in New Orleans)
New Orleans to New York on the *Crescent* train: 30hr

### LOS ANGELES TO MIAMI
Los Angeles to New Orleans on the Sunset Limited train: 45hr (9hr 20m overnight stop in New Orleans)
New Orleans to Greensboro on the *Crescent* train: 19hr 45m
Greensboro to Cary on the *Carolinian* or *Piedmont*: 1hr 30m
Cary to Miami on the *Silver Star*: 4½hr

© KARSTEN MAY / GETTY IMAGES

was so damply warm in the Straits of Malacca that for the first time during my trip I confessed myself uncomfortably hot', she wrote. 'It was sultry and foggy and so damp that everything rusted, even the keys in one's pockets, and the mirrors were so sweaty that they ceased to reflect.'

## A CLOSE CALL IN SINGAPORE

A further delay reaching Singapore made things perilously tight for Bly to make her next connection in Hong Kong. She passed her day in Singapore fretfully, exploring markets crowded with caged monkeys (Bly purchased one as a travelling companion) and observing a local funeral where Chinese lanterns, roast pigs on spears and an army of mourners in white satin accompanied the deceased on their journey to the afterlife. The *Oriental* sailed out of Singapore on 18 December, but Bly's relief at being able to continue on towards China was short-lived.

Monsoon winds churned the South China Sea into a watery mountain range. 'The terrible swell of the sea during the Monsoon was the most beautiful thing I ever saw', she wrote. 'I would sit breathless on deck watching the bow of the ship standing upright on a wave then dash headlong down as if intending to carry us to the bottom.' Bouts of seasickness and injuries from being thrown against the railings marked the journey north to the calm shelter of Hong Kong harbour.

Above: Bly in 1888, before embarking on her circumnavigation. Left: 19th-century Singapore.

## STUCK IN HONG KONG

Despite the tumultuous seas, the *Oriental* arrived in Hong Kong two days early, significantly improving Bly's mood. A cannon was fired to mark the ship's arrival into a bay already crowded with 'heavy iron-clads, torpedo boats, mail steamers, Portuguese *lorchas*, Chinese junks and *sampans*.' There was little time to explore, however; Bly charged to the offices of the Oriental and Occidental Steamship Company to secure her passage to Japan.

It was here that Bly found out about the existence of her rival, Elizabeth Bisland, and the competing attempt to circumnavigate the globe in the opposite direction. Worse still, Bisland's editors had pumped her with funds to bribe ships' captains to leave early, giving her a three-day lead, while Bly would be delayed in Hong Kong for another five days. The mission to circle the globe in fewer than 80 days appeared doomed.

## OBSERVATIONS ON THE EAST

Stoically, Bly accepted her predicament, throwing herself into Hong Kong life in the company of the captain of the *Oceanic*, the ship that would bear her on to Yokohama and San Francisco. She wandered around swanky Happy Valley, home to Hong Kong's colonial elite, learned rudimentary pidgin to communicate with local people, and rode the steam-powered Peak Tram part of the way up Victoria Peak, ascending finally in a sedan chair. The view impressed Bly. 'One seems to be suspended between two heavens', she wrote. 'Every one of the several thousand boats and *sampans* carries a light after dark. This, with the lights on the roads and in the houses, seems to be a sky more filled with stars than the one above.'

Not content with observing colonial life, Bly travelled to Canton (now Guăngzhōu) on Christmas Eve, on a steamer crowded with opium smokers. She spent Christmas Day clattering through the winding lanes of Canton in a sedan chair, dropping in on the US Consul and – somewhat incongruously – visiting the city's execution ground, stacked with barrels

© PRINT COLLECTOR / GETTY IMAGES

of human heads and instruments of torture. Her writing about the punishments meted out to Chinese criminals played a major role in perceptions of China back home in the US.

At last, the *Oceanic* steamed out of Hong

Left: curious locals in 19th-century Canton (now Guăngzhōu), China. Below: Queen's Road in Hong Kong, China, around 1870.

© ALINARI / GETTY IMAGES

Kong harbour for Yokomaha and Bly was able to resume her journey, passing New Year's Eve somewhere out in the Philippine Sea. The 120 hours the writer spent in Japan were some of the happiest of her trip. 'If I loved and married,' she wrote, 'I would say to my mate: "Come, I know where Eden is" and like Edwin Arnold, desert the land of my birth for Japan, the land of love–beauty–poetry–cleanliness.'

## FIVE DAYS IN JAPAN

In her five days in Japan, Bly visited a geisha house, attended a cremation, paid her respects to the 13m-tall (42ft) Daibutsu Buddha statue at Kamakura, and took lunch with the crew of the USS *Omaha* in Yokohama harbour. But her inner traveller was itching to continue the journey, and it was with great enthusiasm that Bly boarded the *Oceanic* for the crossing to San Francisco on 7 January 1890.

The voyage was far from easy. Storms delayed Bly's progress across the Pacific and fellow passengers came to the opinion that her monkey was a bad omen and threatened to throw the poor beast overboard. More mishaps followed. The ship's Bill of Health was temporarily mislaid, threatening a two-week quarantine delay in San Francisco, and reports arrived of snowstorms blocking rail transport across America.

## RETURNING HOME

In the end, however, the *Oceanic* reached San Francisco on schedule, and the editor of the *New York World* chartered a private steam train to carry Bly all the way across America to the East Coast. The journalist was greeted as a celebrity at every stop along the route, with large crowds gathering to catch sight of the famous adventurer.

She wrote: 'I only remember my trip across the continent as one maze of happy greetings, happy wishes, congratulating telegrams, fruit, flowers, loud cheers, wild hurrahs, rapid hand-shaking and a beautiful car filled with fragrant flowers attached to a swift engine that was tearing like mad through flower-dotted valley and over snow-tipped mountain.'

## THE END OF THE LINE

Finally, on 25 January 1890, at 3.51pm, Bly steamed into New Jersey, 72 days and six hours after leaving the quay at Hoboken – circumnavigating the world in record time, four-and-a-half days faster than Elizabeth Bisland's journey in the opposite direction. Bly's record stood for a grand total of four months – shipping magnate George Francis Train circled the planet in 67½ days later that year. But her place in the history books was assured.

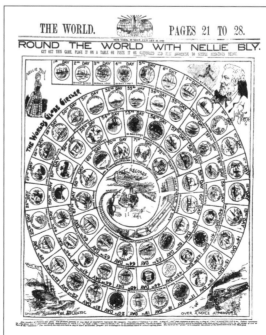

Above: the path through Uyeno Park to Kamakura's Daibutsu (big Buddha), which Bly visited during her five days in Japan. Left: the Daibutsu. Below: the 'Round the World with Nellie Bly' game, one of many promotional stunts that celebrated Bly's 72-day odyssey.

243

# THE DASH TO NEW YORK

**In her 1890 travelogue, Around the World in Seventy-Two Days, Nellie Bly recalls her rapturous reception in California and 'running like hell' on the Southern Pacific.**

I only remember my trip across the continent as one maze of happy greetings, happy wishes, congratulating telegrams, fruit, flowers, loud cheers, wild hurrahs, rapid hand-shaking and a beautiful car filled with fragrant flowers attached to a swift engine that was tearing like mad through flower-dotted valley and over snow-tipped mountain, on—on—on! It was glorious! A ride worthy [of] a queen. They say no man or woman in America ever received ovations like those given me during my flying trip across the continent. The Americans turned out to do honor to an American girl who had been the first to make a record of a flying trip around the world, and I rejoiced with them that it was an American girl who had done it. It seemed as if my greatest success was the personal interest of every one who greeted me. They were all so kind and as anxious that I should finish the trip in time as if their personal reputations were at stake.

The special train had been waiting for my arrival in readiness to start the moment I boarded it. The Deputy Collector of the port of San Francisco, the Inspector of Customs, the Quarantine Officer and the Superintendent of the O. and O. steamers sat up all the night preceding my arrival, so there should be no delay in my transfer from the *Oceanic* to the special train. Nor were they the only ones to wait for me. One poor little newspaper woman did not see bed that night so anxious was she for an interview which she did not get. I was so entirely ignorant about what was to be done with me on landing, that I thought I was someone's guest until I was many miles away from San Francisco. Had I known in advance the special train was mine, every newspaper man and woman who cared to should have been my guest.

My train consisted of one handsome sleeping-car, the San Lorenzo, and the engine, The Queen, was one of the fastest on the Southern Pacific.

'What time do you want to reach New York, Miss Bly?' Mr. Bissell, General Passenger Agent of the Atlantic and Pacific system, asked me.

'Not later than Saturday evening,' I said, never thinking they could get me there in that time.

'Very well, we will put you there on time,' he said quietly, and I rested satisfied that he would keep his word.

It did not seem long after we left Oakland Mole until we reached the great San Joaquin valley, a level green plain through which the railroad track ran for probably three hundred miles as straight as a sunbeam. The road-bed was so perfect that though we were traveling a mile a minute the car was as easy as if it were traveling over a bed of velvet.

At Merced, our second stop, I saw a great crowd of people dressed in their best Sunday clothes gathered about the station. I supposed they were having a picnic and made some such remark, to be told in reply that the people had come there to see me. Amazed at this information I got up, in answer to calls for me, and went out on the back platform. A loud cheer, which almost frightened me to death, greeted my appearance and the band began to play *By Nellie's Blue Eyes*. A large tray of fruit and candy and nuts, the tribute of a dear little newsboy, was passed to me, for which I was more grateful than had it been the gift of a king.

We started on again, and the three of us on the train had nothing to do but admire the beautiful country through which we were passing as swiftly as cloud along the sky, to read, or count telegraph poles, or pamper and pet the monkey. I felt little inclination to do anything but to sit quietly and rest, bodily and mentally. There was nothing left for me to do now. I could hurry nothing, I could change nothing; I could only sit and wait until the train landed me at the end of my journey. I enjoyed the rapid motion of the train so much that I dreaded to think of the end. At Fresno, the next station, the town turned out to do me honor, and I was the happy recipient of exquisite fruits, wines and flowers, all the product of Fresno County, California.

The men who spoke to me were interested in my sun-burnt nose, the delays I had experienced, the number of miles I had traveled. The women wanted to examine my one dress in which I had traveled around, the cloak and cap I had worn, were anxious to know what was in the bag, and all about the monkey.

While we were doing some fine running the first day, I heard the whistle blow wildly, and then I felt the train strike something. Brakes were put on, and we went out to see what had occurred. It was hailing just then, and we saw two men coming up the track. The conductor came back to tell us that we had struck a hand-car, and pointed to a piece of twisted iron and a bit of splintered board—all that remained of it—laying alongside. When the men came up, one remarked, with a mingled expression of wonder and disgust upon his face:

'Well, you ARE running like h–!'

'Thank you; I am glad to hear it,' I said, and then we all laughed. I inquired if they had been hurt; they assured me not, and good humor being restored all around, we said good-bye, the engineer pulled the lever, and we were off again. At one station where we stopped there was a large crowd, and when I appeared on the platform, one yell went up from them. There was one man on the outskirts of the crowd who shouted:

'Nellie Bly, I must get up close to you!'

The crowd evidently felt as much curiosity as I did about the man's object, for they made a way and he came up to the platform.

'Nellie Bly, you must touch my hand,' he said, excitedly. Anything to please the man. I reached over and touched his hand, and then he shouted:

'Now you will be successful. I have in my hand the left hind foot of a rabbit!'

Well, I don't know anything about the left hind foot of a rabbit, but when I knew that my train had run safely across a bridge which was held in place only by jack-screws, and which fell the moment we were across; and when I heard that in another place the engine had just switched off from us when it lost a wheel, then I thought of the left hind foot of a rabbit, and wondered if there was anything in it.

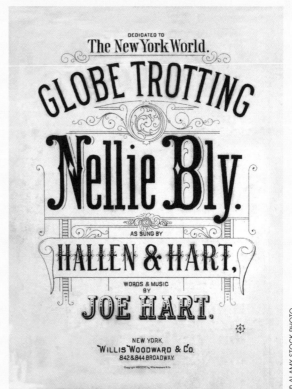

# NEXT STOP: 1000KM/H

**From steam and diesel to 'maglev' and 'hyperloop', rail travel has always been the most successful way to transport the greatest numbers as fast as possible.**

When Nellie Bly embarked on her ambitious journey around the world, steam was the only game in town. Trains ran on it. Ships ran on it. Even factories ran on it. Karl Benz only launched his Patent Motor Car Model 1 in 1885, and it looked more like a perambulator than a serious means of conveyance. The running speed of a steam train when Nellie Bly embarked from Hoboken was around 65km/h (40mph) – dizzyingly fast by the standards of the time, but a snail's pace compared to the trains of today.

Early attempts to push trains faster often ended in disaster, with hundreds killed in accidents caused by ineffective brakes. It was only with the introduction of continuous automatic braking that the full power of steam could be unleashed. But steam still had antisocial side effects: deafening noise, teeth-jarring vibration and smog, which transformed Victorian-era cities into choking smokehouses.

The petroleum-powered motor car went mainstream with the launch of the Model T Ford in 1908, but it took a further four years for the first diesel-powered locomotive to roll on to the Winterthur–Romanshorn railroad in Switzerland. For many rail companies, electricity looked like a more promising proposition – 4 miles of the Baltimore and Ohio Railroad were electrified in 1895, and Italy succeeded in electrifying the entire 106km (65-mile) line from Lecco to Valchiavenna in 1902.

The US, however, caught the oil bug hard, and General Electric and Westinghouse's prototype electric trains were abandoned in favour of diesel-electric locos, which used diesel engines to power an electrical generator. The Kaufman Act of 1923 hastened the demise of US steam transport; all steam trains were banned from New York City, and fully electric trains were dropped after the Great Depression in favour of cheaper diesel-electric technology.

A similar revolution was taking place in Europe. Swiss and German engineers created the first diesel-electric trains in 1914, but it wasn't until 1922 that these hybrid railcars were put to work transporting the nation. By 1934, diesel-electric 'streamliners' were hauling passengers on both sides of the Atlantic.

After WWII, the US wasted no time in modernising its railways, leasing its obsolete steam trains to the USSR, and throwing its weight behind diesel-electric technology. European makers hedged their bets, backing both diesel-electric trains and newly invented diesel-mechanical engines. Freed from the engineering problem of controlling power output from a steam-powered engine, trains started to accelerate. In the post-war years, *California Zephyr* passengers travelled at over 160km/h (99mph) on the flat terrain of the central United States.

For the next 20 years, trains were improved incrementally, rather than radically. The prosperous West followed a mix of diesel and electric services, while poorer nations struggled on with ageing steam stock. Japan, however, had loftier ambitions. While most of the world was shunting around on aerodynamically inefficient, snub-nosed trains, Japan created a new type of train that cut through the air like a bullet – the *Tōkaidō Shinkansen*.

The first high-speed journey from Tokyo to Shin-Osaka in 1964 took four hours. Pushing the boundaries of aerodynamics and electric propulsion, Japanese engineers reduced the journey time to 2hr 30m by the 1990s.

But propelling a train at great speed in contact with tracks created huge amounts of heat from friction, posing a serious challenge to braking. Enter Birmingham, in the English West Midlands. In 1984, Birmingham Airport unveiled a 600m (1970ft) train line where carriages levitated magnetically 15mm (0.6in) above the rails – the first commercial maglev (derived from 'magnetic levitation') transport system.

Germany briefly experimented with the same technology in 1991, but it was engineers in Asia who pushed things forward. Combining maglev science with *shinkansen* know-how, China unveiled the world's fastest train – the 431km/h (268mph) Shanghai Transrapid – in 2004. Today, Japan and South Korea are vying to be the first to run maglev trains at over 600km/h (370mph), while America's pneumatic Hyperloop has its sights on the 1000km/h (620mph) barrier.

# BY
# MOTORCYCLE

**M**otorcycles and travel have gone together since 1885 when Gottlieb Daimler's son Paul rode the world's first internal-combustion-powered motorcycle, the Reitwagen, on its admittedly short inaugural journey.

Indeed, Carl Stearns Clancy on his brand-new 1912 Henderson motorcycle preceded the first successful attempt around the world by car by an impressive 16 years. And if that's not enough, the record for the fastest journey around the world by land, at 19 days and four hours, is also held by a motorcyclist. Alone on a Yamaha sports bike, Nick Sanders beat the fastest car, even though it was allowed two drivers, by nearly two days.

There have been many RTW riders since Carl Stearns Clancy (RTW is motorcyclists' acronym for 'round the world'). The best-known are probably Robert Edison Fulton, who wrote a book and made a film about his journey in the early '30s, and Ted Simon, who rode over 100,000km (62,100 miles) to complete his first circumnavigation by motorcycle between 1973 and 1977, and repeated the trip between 2001 and 2003.

The first woman to circumnavigate the world solo on a motorcycle was Anne-France Dautheville, in 1973. She acknowledged the dangers but also recounted the story of a flat tyre in Afghanistan, where three men helped her. One made her sit down, the second stood in front of her so she had shade and the third repaired the tyre. Seven years later, Elspeth Beard (right) became the first British woman to ride around the world.

One thing that makes RTW motorcycle trips attractive is the relatively low cost and limited amount of preparation that's required. Find the right bike, equip it from a catalogue, sort the paperwork and passport, get your shots and have them noted in your vaccination book – that's it, you're pretty much ready to go RTW. Since Clancy's day, motorcycle technology has changed almost beyond recognition. Today a plethora of so-called 'adventure bikes' is available, offering taller suspension to cope with poor or non-existent roads, large fuel tanks to enable them to cover long distances between fuel stops, strong construction and secure luggage. But the essentials remain the same: two wheels, an engine, a way of carrying gear and somewhere – hopefully reasonably comfortable – for the rider to sit.

Adventure motorcycles as such became available in the early 1980s, led by BMW's R80G/S (Gelände/Strasse, or terrain/road). The initials later came to stand for Gelände/Sport, in an effort to emphasise the fun nature of the bikes. Almost all other motorcycle manufacturers have followed BMW into this 'Gelände' as it has proven to be highly popular. Many people would like to think of themselves as RTW riders, but will never find the time. Buying an adventure bike is the next best thing.

RTW riding is remarkably affordable once you have the bike, and really you can ride just about anything. While a 'proper' adventure bike is good, it's not necessary unless you want to tackle truly difficult stretches such as Siberia's Road of Bones. Eastern Russia is one of the few parts of the world to be impassable for motorcycles for much of the year, with its deep mud or bone-chilling cold. In Africa it depends on the actual route you choose. The only completely untraversable stretch is the Darien Gap connecting Central and South America – otherwise, saddle up.

# POPPING OUT FOR A GUINNESS...

**If you're going to ride from Australia to Dublin, you might as well keep going. Peter Thoeming remembers how an idyllic RTW gave him the long-distance-riding bug.**

'You are very fortunate,' said the manager of our hotel in Satna to my travelling companion Charlie and me. He had joined us at breakfast and when we looked at him quizzically over our teacups, he explained. 'You speak English. Wherever you go in the universe, you will find someone who speaks English.'

That turned out to be pretty well true on our around-the-world motorcycle trip, although of course not everyone understood the language, and sometimes it paid not to shout about our proficiency. An army officer inspecting our passports a few weeks later at a roadblock near the Afghan capital, Kabul, looked at the covers and said 'Australia. So, you don't speak English?' We both shook our heads solemnly and he waved us through.

Besides our ability to both speak English and deny it, my friend Charlie and I were lucky in other ways. We also had the time and the means to tackle that trip. Motorcycle travel is an interesting activity at the worst of times, and at the best – when everything, including fascinating cultures, riveting scenery, unfamiliar food and unexpected encounters, falls into place – it is wonderful. Over drinks late one night in 1977, we had come up with the idea of visiting the Guinness Brewery in Dublin, and that seemed to naturally segue into a round-the-world ride.

Planning the trip was easy. We would drift from east to west, stopping to look at anything interesting along the way and traversing countries which would give us visas. We'd try to avoid the monsoon and ride in dry seasons as much as possible.

Clearly we would need secure luggage, spare parts and tools, comfortable and durable clothing and a first aid kit. I would strap my camera bag on top of my bike's luggage rack. We arranged all of this in stages, starting with light, small-capacity Honda XL250 dirt bikes, which had reasonably long suspension. We added plastic long-range tanks and had steel cages welded up for luggage and extra fuel and water tanks. Everything was secured with a handful of small, keyed-alike padlocks.

We set off in February 1978 were on the road for two years, eight months. On our trip we met a number of people who, like us, were doing what they could with the transport available to them. A Dutch chap was travelling on an ex-WWII Harley-Davidson WLA with a sidecar, ponderous but convenient. Unfortunately it was too heavy to transport by air, which restricted his choice of destinations. We had deliberately chosen the small, light XL250s so that we could fly across Burma (as it was then), which was officially impassable to road traffic.

Another chap had started out on a new Honda Four, a 750cc sport bike that simply would not run on the poor-quality fuel east of Iran. He stored it in Pakistan and continued as a backpacker. Yet another, whom we met on his way back, was on a Yamaha 125cc dirt bike which he'd ridden from Calcutta (now Kolkata) to Kayseri in Turkey. We also encountered a French brother and sister who had pedalled a pair of Velo Solex motorised bicycles from France to Pakistan where they had sold the totally worn-out bikes.

Transporting the little Hondas where we couldn't ride them proved to be quite easy. We rode from Sydney to Perth, or rather Fremantle, and shipped them from there to Singapore. In those days a combination freighter/passenger

Opposite: Peter in Oregon, USA, celebrating his first sighting of the Pacific Ocean; Charlie checks a wobbly wheel in Thailand.

ship plied this route at a reasonable cost, and we could travel with the bikes, which found a snug home in the hold. Today a RTW rider would have to consign their motorcycle to the belly of an aircraft. We had to do that only from Bangkok to Kathmandu in Nepal, and once again across the Atlantic.

Roads around the world vary as much as people do. From the huge mudslides across Himalayan roads, permanent enough to have names, to the 'super slab' interstates of the US, and from the potholes of Turkey to the muddy dirt tracks of backcountry Macedonia and gravel roadbeds of remote Australia, they all present their own challenges. Fortunately, light bikes and basic caution made things pretty easy for us. It's different if you're tackling Siberia – in winter or in summer.

The most difficult route I have faced was the crossing of the Sahara, undertaken with my wife-to-be, Anne, whom I had met on the ship from Fremantle to Singapore. Charlie had headed off to complete his RTW ride, but I had decided to take a look at North Africa first. The Sahara is actually not bad, except for huge sand drifts which can be challenging on a motorcycle. What made it tough was the emptiness. Even for an Australian, the Sahara's vast landscape is oppressive. As I rode along, I imagined that the engine was rattling, that a tyre was going flat, that the bike was breaking down from the poor-quality fuel... It was all in my mind, but nerve-racking nonetheless. A vicious sandstorm just out of El Goléa (now El Menía, Algeria) made it clear to me that the dangers weren't all in my head, after all.

Talking of dangers, and protecting against them, we set out with a fairly blasé attitude to paperwork, without worrying particularly about visas. The one thing we did take seriously

was vaccination, especially against cholera. We carried both serum and hypodermics and found a clinic in Kabul to administer the shots and stamp our books. The doctor was delighted when we told him that he could keep the syringes. Everything, including hypodermic needles, got reused in places like Afghanistan.

Visas were little trouble; we just applied for them at the last major town or city before the border, or even at the border. That backfired on Anne and me once. At the Tunisian border, the guard sucked his teeth in the way that often indicates trouble to come. 'Where is your visa?' he asked. 'The Tunisian Tourist office in London told me we did not need visas', I said. 'Ah, there is the problem. It is their job to get people to come to Tunisia. It is my job to keep them out.'

Our Algerian visa was single-entry, so we couldn't go back. We spent an eventful night at the border post – they were worried about incursions from Libya – but we were given food by kindly guards and let go the next morning.

When we rolled off the ferry at Dover on arriving in England, the immigration official turned the pages of our passports and looked at all the strange stamps with pursed lips. Then he glanced up and said: 'Not carrying any banned substances, are we?' When we assured him that we weren't, he waved us through.

I still had the USA in front of me, but, as I had expected, it proved to be smooth riding among wonderfully friendly folk. My six weeks zigzagging cross the country passed quickly and Honda America was so impressed with my ride that they offered to ship the bike back to Australia from Los Angeles, which saved me a lot of money I no longer had.

I planned my next motorbike trip on the flight across the Pacific. This long-distance riding is addictive.

# THE MAN WHO RODE HIS LUCK

**Before Robert Edison Fulton knew it, a careless remark in 1932 had him on the road aboard a top-of-the-range Douglas – and around the world on a trip where he caught every break.**

At a dinner party in London in 1932, Robert Edison Fulton Jr, was asked whether he would soon be sailing home to New York. For reasons he was never able to recall, he replied, 'Oh no. I am going to ride around the world on a motorcycle.' One of the owners of the Douglas Motorcycle Company happened to overhear him and before he knew it, he was hitting the road equipped with a customised bike, a movie camera and 4000ft of film.

## MOTORCYCLE STUDIES

The initial plan of the architecture student, who had spent a year on a postgraduate course at the University of Vienna, was to motorcycle his way to see the architectural delights of Europe, the Middle East and Asia. Fulton had started exploring early – as a young boy, he had been on board the world's first commercial airline flight from Miami to Havana and travelled with his family to Egypt when Tutankhamun's tomb was opened – and a spirit of curiosity and adventure seemed to run through his veins. Lightbulb inventor Thomas Edison may have been a family friend, rather than a blood relative (and neither was steamboat pioneer Robert Fulton a known relation), but Fulton showed promise as an inventor himself – he built his own car in high school, and in later life invented the Airphibian, a flying car, and an aerial rescue system, Skyhook. Fortunately, he also had some ability as a mechanic; the 6HP, twin-cylinder Douglas motorcycle he rode required frequent maintenance.

Fulton ended up covering 40,200km (25,000 miles) over 18 months, and went on to write about his experience in the *One Man Caravan* (published 1937). Many years later, with the help of his two sons, he produced two films based on his epic ride, and also exhibited a selection of stills from the reels shot on the road.

Despite crashing the only other motorcycle he had ever owned after only a few rides, Fulton quickly took to the Douglas – although the sealed roads of Europe bored him. There were bumpier rides and other obstacles ahead.

He waited six hours at the Yugoslavian frontier, then a whole day crossing into Bulgaria (where he finally took matters into his own hands and rode through without permission) and three days to get into Greece.

In Turkey, Fulton first encountered the Asian night 'with its silent, dark and starlit mystery', interrupted by the cacophony of the caravans plying the same route. Other memorable events included staying the night in a Turkish jail as a boarder, having to pay for his bed, and riding off an unfinished bridge and crashing into the stony riverbed below. When Fulton awoke after this accident, he was in a mud-room with his motorcycle, being tended to by locals.

From the 4.5m (15ft) drop of the bridge, the motorcycle had suffered only a slightly bent front fork, giving it a tendency to pull to one side. Also, all the oil had seeped out of the tank, but the locals came to his aid again, supplying him with mustard oil to use as a substitute. Otherwise the bike was fine, as was Fulton.

He rode on through Syria, Lebanon and into Iraq, finding helpful if bemused locals and expatriate members of the French and then British administrations wherever he went. Following the armoured cars of the French dawn patrol, which served to protect the track from Damascus to the border with Iraq – at least until the dust in their wake made it impossible – he began the crossing of the desert to Baghdad, meeting a workmate of his father's along the way. .

## SKIING IN KASHMIR

Struck down with jaundice, Fulton spent seven weeks in the British hospital in Baghdad and then, still in poor shape, continued south to Basra on the Persian Gulf by train. Trying to console himself for missing 'Shiraz and Persepolis, Teheran and Meshed', he acknowledged that he needed to pause in order to fully recover. A tramp steamer voyage onwards to Karachi granted him some recuperation time.

In 1933, Karachi was still part of India – Partition was another 14 years away – and Fulton noted that the density of the Indian traffic meant that hardly anyone even

Robert Edison Fulton, photographed in 2000 on the restored Douglas Motorcycle Company bike that he rode around the world in 1932–33.

From left: one of Fulton's FA-2 Airphibians flies above Manhattan in 1947; a hybrid of plane and car, the Airphibian had a removable propeller and wings; Fulton in the cockpit of an Airphibian with his wife, Florence, in 1949; a family send-off as Fulton takes flight from his Connecticut garden in 1949; the Airphibian makes it to Broadway.

noticed his motorcycle. He made use of the government-run dak bungalows, set up for the convenience of travelling magistrates, inspectors and the postal service, and the waving canvas flaps of their human-powered air-conditioning. After crossing the rough, wild-boar-infested countryside to Udaipur, he met the Maharana's stable-master. For three weeks, he helped exercise the ruler's horses with his new acquaintance, who also managed to explain the Indian caste system to him.

In Lahore, as Fulton admired an enamelled Automobile Association of Northern India sign with the legend 'LONDON – 6,372 MILES', he was struck by a car. Once again, thankfully, he was spared any injuries. The crash bent the bike's swingarm – which compensated exactly for the bent front fork it had suffered in Turkey. It ran straight again. The offending car sustained a broken front axle.

Next, accepting an invitation to go skiing from a Royal Air Force officer in Delhi, Fulton rode up into Kashmir along a road that was subject to such frequent rock slides that the State of Kashmir had to maintain a huge staff in readiness to rebuild the road. He parked his motorcycle and took a native guide, but still became lost in deep snow on the trek to the remote ski field and nearly perished. He enjoyed the skiing, though.

Heading west into what is now Pakistan, Fulton managed to pass through the checkpoint designed to keep practically everyone out of the Tribal District of Waziristan because he was wearing a pith helmet in the same pattern as the ones worn by British officers.

Riding on along roads bustling with military convoys, owing to trouble stirring on the North-West Frontier Province (now Pakistan's Khyber Pakhtunkhwa), Fulton nevertheless reached Peshawar. Here he was struck by the melancholy thought that the climax of his trip was already past. So, when he was strolling in Peshawar and noticed a brass plaque with 'Royal Afghan Consulate' engraved upon it, he didn't hesitate to enter the building and request permission to visit.

## INTO AFGHANISTAN

Once again luck was on his side. After passing through the hands of innumerable flunkeys, he finally reached the Consul General who, to Fulton's delight, was said to speak German, which Fulton also spoke. Never slow to take advantage, Fulton asked not only if he could get an Afghan visa but whether he could take pictures. The plenipotentiary steadily replied, 'Ja, ja.' Neither Fulton nor the flunkeys could quite believe it, but as his papers were being stamped, Fulton heard the explanation. His Excellency, it seemed, had been in Germany a long time ago and the only word he still remembered in that language was... ja.

Fulton was fascinated by Afghanistan – especially by the houses, or more accurately small castles, in which the Afghans lived. At one point he thought he might have found the remainder of the Ten Lost Tribes of Israel, in the tribe of the Abdalis, who call themselves the children of Israel. As he turned south towards Kandahar, his progress was delayed by the many fords in the roads, which he had to check

on foot before braving them astride his bike. After crossing back into Pakistan near Quetta, he headed south to the sea.

An overnight journey down the Arabian Sea on a freighter took Fulton to Bombay, where he had his exposed film developed and shipped home, before hurtling along the Grand Trunk Road to Calcutta. There, he convinced an agent to sell him a deck passage to Sumatra. By now, Fulton was running out of money, and wouldn't have access to additional funds until Singapore. Despite low reserves, he enjoyed Indonesia, then still in Dutch hands, Singapore and Malaya. He was also impressed by the electric railways in Thailand and rode the tracks on his motorcycle when the roads were impassable. The signalling system had been adopted from America, so he knew when to 'jump the rails'.

## SHANGHAI EXPRESS

Fulton rode north through what was then Indochina into China proper, and found it intriguing. He rode upcountry, where he found women with their feet still bound and constant tales of the depredations of bandits, whom he mercifully never encountered. Then the rains began and, for once, Fulton recognised when he was beaten. He retraced his route by railway into Indochina to the coast at Haiphong, where – once more out of money – he put up the Douglas as security for a berth on a ship bound for Shanghai, where funds awaited him.

Fulton rode north from Shanghai despite being told that he needed a Chinese licence plate, which was apparently unobtainable. He bluffed his way through the inevitable roadblocks

## THE CAR THAT FLEW

In the 1940s, Robert Edison Fulton anticipated the optimistic spirit of the age if not quite its means of transport. It took five minutes to remove the propeller and wings from his Airphibian, to convert it to a car. On the road, it reached 80km/h, in the air 175km/h. The US' Civil Aeronautics Administration even licensed it. Fulton imagined an Airphibian in every suburban garage. But spiralling development costs killed his flying car dream.

with business cards he had had printed in Shanghai. They featured his name in Chinese script and identified him as a 'scholar' to the border staff, who happily let him pass.

China's cities, the Great Wall and the Grand Canal all left a strong impression on Fulton. So did the food. He narrowly avoided being served monkey stuffed with rice while it was still alive and then roasted, and restricted himself to a diet of boiled rice for a while. In Sianfu (now Xi'an) he mellowed and partook of, in his words, 'lumps of flesh boiled in oil with a taste of sweet and sour' as well as 'assorted cakes' – he came to regret that, less than an hour out of town.

Japan presented another surprise. On his arrival, Fulton was met at the pier by newspaper reporters – he was known. The American was even escorted by 33 members of a motorcycle club at one stage. The club's spokesman assured him that 'Japan and America will come closer together through the motorcycles.' Although he found Japan thoroughly modern, Fulton was initially fazed by the mixed bathing in its inns.

The idea of writing a book about his journey was first presented to Fulton in Tokyo while at tea with the American ambassador. Once aboard the transpacific liner that carried him and the Douglas to San Francisco, he had time to mull things over and the story of his ride, *One Man Caravan*, was born.

## BACK FOR CHRISTMAS

Back in the US, only the ride across from San Francisco to his family home in New York remained. He 'planned to scoot three thousand miles swift as the wind' in order to be home for the Christmas holiday. One night in Dallas, the Douglas was stolen from beside Fulton's cabin. Fortune favoured the adventurer once again – his motorbike was found within a week, and he completed his ride. A day before Christmas, and seventeen months after starting his trip, he arrived at his parents' home in New York.

Fulton went on to work as an aerial photographer for Pan American World Airways later in the 1930s and subsequently used his photographic and mechanical skills to design an aircraft gunnery simulator for the US Navy. After teaching himself to fly his wife's aircraft, he designed and built the Airphibian (see panel left), his flying car. During the 1950s he created the Skyhook, a mechanism for safely plucking people from the ground into an aeroplane. The system was used by, among others, James Bond in *Thunderball*.

Fulton died in 2004 at the age of 95 and is remembered, according to a eulogy in *One Man Caravan*, 'as a kind, wise, generous and intensely interesting man'.

# WOMAN ON A MISSION

**Harrassment, condescension, and even a miscarriage – four decades ago, Elspeth Beard contended with them all to become the first Englishwoman to ride around the world.**

Not until 2008, 24 years after she had motorcycled her way around the world, did Elspeth Beard discover that she was the first British woman to have done so. 'It felt very odd and still does feel strange,' she told Lonely Planet.

It's hard to avoid the conclusion that the long indifference to her achievement derives from the fact that Beard is a woman. Which is particularly galling because on her 56,325km (35,000-mile), two-year circumnavigation via the US, Australia, Southeast and South Asia, Turkey and Europe, Beard endured all that a man would have faced, and, as a young woman, much more – including a miscarriage, and frequent sexual harassment suffered in upmarket homes in Sydney as well as hostels in India.

But she is no victim. As is clear from Beard's 2017 account, *Lone Rider: The First British Woman to Motorcycle Around the World* (Michael O'Mara), she began her trip in 1982 as a single-minded, well-organised and trusting 23-year-old. Two years and one circumnavigation later, she rode into London early one morning, tough, wily and clear-eyed.

## A TOUGH START

The young architecture student's plan was to ride across the USA, get to Australia, then work for a year to save money for the remainder of the trip. Beard freely admits that, apart from that sketchy itinerary, she had little idea of the nuts and bolts of accomplishing such an epic journey on a motorbike. Her attempts to gain sponsorship and media coverage failed. She did at least have some long-distance miles on the clock: with her brother Justin, Beard rode across the US in 1981. It was then that she had idly considered a round-the-world motorcycle trip.

What determined Beard to make her idea reality was heartbreak: a big relationship had ended suddenly. So she worked in a central London pub to save money while she completed, without distinction, the first part of her architecture studies. She outlined her plans to her friends and family; her parents were dismayed. Nonetheless, in September 1982, she put her 1974 BMW R60/6 on a cargo ship bound for New York, and a few weeks later found herself at Heathrow waiting for a transatlantic standby flight, almost as confused and upset as her mother about her imminent trip.

Beard was reunited with her bike in New York and, noting her mileage (45,477 miles, 73,188km), off she rode, first for Detroit, to stay with an aunt, then to Los Angeles via Tennessee, New Orleans and Texas. Occasionally, she confesses in her book, swathes of America passed in a blur as she struggled to shake off thoughts of her failed relationship, fended off sleazy men, and fretted about the cost of motels and securing an Australian work permit. But at last the journey was under way.

## DISASTER IN NEW ZEALAND

Because of the work permit problem, Beard was forced to make a detour via New Zealand. There, one night in a youth hostel, she awoke in the top bunk, in pain: 'I was lying in a pool of blood'. In the most distressing way, a short relationship with an engineer called Mark in the months before her departure from London had caught up with her. Unaware that she was even pregnant, she had miscarried. On top of the physical and emotional pain of that experience, Beard struggled with having to break the news to Mark, with whom she did not want a relationship. To add to her predicament, he had recently written to say he was relocating from London to Australia to be near her. Yet she recovered, rallied and secured her work permit. Finally, Beard and her bike arrived in Sydney.

Maintenance on the road in India.

There, she caught up with friends, pondered what sort of relationship she wanted with Mark, and worked at two architecture practices. One employer refused to give her anything but the most menial work; the other tried to seduce her at his home. The latter, though, did at least assign her a project designing 270 new houses.

Meanwhile, Beard remained focused on funding the next stage of her trip. To reduce her overheads, she moved into a garage with her bike, drastically cut her hair (to save on shampoo!) and prepared her motorcycle for the ride across the continent to Perth, the very long way: north up the coast, west to the Northern Territory, south via Ayers Rock (as Uluru was then called) to Port Augusta and west across the Nullarbor Plain.

## BEATING THE OUTBACK

It was on this leg that the riding became rough, anticipating what lay ahead in Asia. So-called 'road trains', long-haul trucks towing up to three trailers, forced Beard to take evasive action more than once. At one point she rode 805km (500 miles) in a single day on dirt tracks and riverbeds, in 40°C, alone, to the city of Mount Isa. Then she had an accident so serious that she woke up in a Northern Territory hospital with no recollection of what had happened (she had ridden into, but not out of, a large pothole). Days after she was back in the saddle, 15cm (6in) of rain fell in one night, engulfing the roads in impassable mud. If this was adventure riding, Beard was up to her knees in it.

'The cliché of cruising into the sunset simply didn't exist,' she reflects in *Lone Rider*. 'Waypoints needed to be reached, schedules met, storms avoided, although I came to realise that most deadlines were no more than constraints of my own making.' There were highlights: a night by a watering hole under a star-spangled sky in the Outback; a crucial lift with Des, a gnarled old road-train trucker and 'a true gentleman'. Apparent throughout *Lone*

Clockwise, from top: a break from the bike in Iran; the Lone Rider's BMW R60/6 at White Sands, New Mexico; Elspeth Beard on the road to Srinagar in Ladakh.

© ELSPETH BEARD

## PIONEERS OF FEMALE ADVENTURE MOTORCYCLING

### THERESA WALLACH & FLORENCE BLENKIRON

In December 1934 Theresa Wallach (below left) and Florence Blenkiron (below right) left the Aldwych, central London, on their unprecedented ride from London to Cape Town. Lady Astor, the first female British MP to take her seat, told the gathered crowds: 'I am an unrepentant feminist and convinced that whatever a man can do, a woman can do too.' As Wallach noted in her account of the expedition, *The Rugged Road*, Lady Astor was wrong in one important respect: in fact, no man had ever attempted the feat. The two Englishwomen were noted racers and Wallach an engineering graduate. The pair prepared for the expedition for a year and attracted sponsorship. A test run of their 600cc Panther Redwing on Bagshot Heath, Surrey, convinced them of its sturdiness for the 21,700km (13,480-mile) trip through Europe and Africa. They rode with a sidecar and trailer, entirely unsupported – south from Algiers across the Sahara, from oasis to oasis, then southeast to Kampala, Nairobi, and south via Victoria Falls and Bulawayo to Cape Town – a journey that took them eight months. Along the way, they encountered the French Foreign Legion, European royalty, many African peoples, from Tuareg to Pygmies, as well as gorillas and lions, and contended with terrain ranging from desert to jungle. At the southern edge of the Sahara they had to completely rebuild their motorcycle engine. On arrival in Cape Town, a drained Wallach chose to return to England by ship; 'Blenk', meanwhile, decided to ride back, procuring a new bike for the return trip. By April 1936 she was back in London.

© IMAGNO / GETTY IMAGES

© ELSPETH BEARD

© ELSPETH BEARD

*Rider* is her growing confidence as a highly competent long-distance motorcyclist.

## A WOMAN'S BEST FRIEND

Beard's adventures in Southeast Asia got off to a quiet start. She decided to ship her bike to Singapore and travel overland with Mark (who'd been making his own way around the region) through Bali and Sumatra, eventually parting from him in Bangkok. From there she rode first into the north of Thailand, undeterred by stern warnings about the bandits of the Golden Triangle. In fact, it was a stray dog that nearly undid her. By now it was April 1984 and Beard was in a race to get back south to Penang, Malaysia, to catch a boat to the subcontinent and reach Kathmandu in Nepal, where she was due to meet her parents in May. With good roads beneath her, she twisted the throttle hard – and hit a dog. What followed was an episode that illustrates that one requirement of travelling independently is knowing when you should cede that independence.

Beard and her bike were swept up by the community of the local village in rural southern Thailand. While she repaired her BMW, the concerned residents insisted on dressing her flesh wounds and provided food and drink as she toiled in the heat to fix a broken gasket.

'Then, as I was loosening a bolt, I felt a slap on my back,' she writes in *Lone Rider*. 'I turned and was greeted by the sight of a young Thai woman with a raised thumb and a broad smile... Something flashed between us. I knew that she wanted to prove to the men standing around...that we were just as capable

Above, from left: in Thailand, where a local community rallied round after Beard's collision with a dog; photocall at Mae Sai, Thailand.

## PIONEERS OF FEMALE ADVENTURE MOTORCYCLING

## BESSIE B. STRINGFIELD

As a black woman, riding a Harley-Davidson around the US in 1930s and '40s took character and courage, especially in the southern states where the Jim Crow laws enforced racial segregation. But Bessie B. Stringfield was not short of either. A mercurial figure, Stringfield told journalist Ann Ferrar (in her book *Hear Me Roar: Women, Motorcycles and the Rapture of the Road*) that she was 16 when she got her first motorcycle, a 1928 Indian Scout. She didn't know how to ride: 'I wrote letters to the Man Upstairs, Jesus Christ and He taught me.'

By 20 she had her first Harley and had embarked on the first of her summer cross-country tours which, Stringfield told Ferrar, she'd plan by chance: 'I'd toss a penny over a map, and wherever it landed, I'd go.' She travelled light: 'I'd sleep on my Harley at gas stations.' Her entry into the American Motorcycle Association Hall of Fame states that her motorcycle tours took her to all 48 lower states. Stringfield recalled working as a stunt rider in carnival shows and as a dispatch rider for the US Army during WWII. After the war she moved to Miami, where she qualified as a nurse and became a familiar and stirring sight on the streets, a black American woman proudly riding the all-American motorcycle.

© ELSPETH BEARD

## BY MOTORCYCLE

India, the easy-going atmosphere of Nepal was a relief, and she arrived in good time to meet her parents. Travel-worn and exhausted, she had to convince the staff of their luxury hotel to allow her admission. More dispiritingly, her mother was uninterested in her journey: 'I'd achieved something that few people had even attempted, but I felt my mother still reacted to me as if nothing had changed,' she writes. Her father, meanwhile, appeared more excited by the bargain rolls of camera film he'd picked up for the holiday.

### A FELLOW TRAVELLER

Back in Kathmandu after a walking trek of the Annapurna Circuit with Mark, Beard finally met a true fellow traveller: a Dutch biker called Robert, also on a BMW, who had ridden from Australia. Beard was thrilled. 'He was the first person I'd met in nearly two years doing a similar overland journey as myself.' Mark left, and Robert and Elspeth decided to ride together. It was from here, making the difficult journey through northern India, Pakistan and Iran, that Beard revealed further her capacity for long-distance riding.

She went out of her way to visit – and was enthralled by – sights in Rajasthan such as the palaces of Jaipur. She and Robert thrashed their bikes hard to get up over the Zojila pass in the Indian Himalaya to reach Leh and Ladakh. And in Kashmir, they fell deeply for one another.

But otherwise, Beard was constantly vexed

© ELSPETH BEARD

as any man, if only given the chance.' Beard recuperated with a family for three nights. Her attempts to pay for their hospitality were politely but firmly declined. On the final evening she realised what she and her host family had been dining on: the dog she had run over.

Arriving in Madras, Beard had her first, and far from last, encounter with mind-bending Indian bureaucracy in arranging to have her bike transported by train to Calcutta. After

Left: the long road ahead in Western Australia. Below: rough riding on the way to Leh in Ladakh, India.

Left: riding out of Kathmandu, Nepal. Below: buying provisions in India. Opposite: Elspeth and Robert riding the Zojila pass over the Indian Himalaya to Leh.

© ELSPETH BEARD

© ELSPETH BEARD

by northern India's pollution, bureaucratic chaos and the plight of its populace – the latter illustrated starkly in *Lone Rider* by the account of a father apparently pushing his young daughter into the path of Beard's speeding motorcycle to compel her to pay compensation. Thankfully, the girl sustained only minor injuries, and Beard reluctantly handed over 20 rupees, at the time about $2. 'I was shocked at the desperation of the father and the lengths he was apparently willing to go to secure a few rupees. I was also surprised at how little I thought of the girl in the coming days. After only a month in India, the country had hardened me.'

Beard was learning fast in other ways. The first of several tense border incidents took place on the Indian border with Pakistan. Having ridden through Punjab at a time of high political tension, she and Robert were turned away from the border and forced to make a 500km (310-mile) ride back to Delhi for a permit. There, three weeks of infuriating official meetings come to nothing.

The stakes were high: their health was not good and they had met a Dutch family in Delhi in similar straits, fast running out of money and with a very ill child – if the authorities were indifferent to that family's predicament, it seemed the pair stood little chance. They rode back to the border and tried bluffing their way through, in the hope that the border officials

were equally out of their depth. It worked. After several more tense encounters, they held their nerve and obtained the required passport stamp: 'Exit Amritsar'.

Home could not come fast enough. Robert and Beard were suffering from hepatitis, Beard was still having to deal with predatory men and even had to resort to disguising herself as a man in the desert of southwestern Pakistan. Her bike was also on its last legs, suffering on roads that dwindled into very rough trails and, at one point in Pakistan, no road at all. But on through Iran and Turkey, she rode.

## HOME AT LAST

On 22 November 1984, Beard pulled up outside her parents' house on Wimpole Street, central London. In the pre-dawn gloom, she noted the mileage: 74,574 (120,015km). Like many adventurers, she found the return to her old life difficult. 'I simply wanted to prove to my family, my friends and most importantly to myself that I could succeed,' she writes in *Lone Rider*.

Her biking friends delighted in her return and listened to her stories politely. The motorcycle press again brushed off her attempts to gain some recognition, or misrepresented her. Her parents' indifference persisted. 'When I was doing the trip [my mother] couldn't deal with the enormity of it,' she told Lonely Planet. 'I often wonder what she would have made of my book. She never saw [my trip] as an

achievement, but neither did I because I was ignored when I got back.' (Her father died suddenly in 1991.)

Before she reached London, Beard had parted happily from Robert after a few days at his family home in the Netherlands. Yet despite attempts to rekindle their road romance, their relationship stalled. Eventually, Robert visited London to tell her that he had met someone else who insisted that he break off contact with Beard. Heartbroken, she agreed, writing: 'I found it almost impossible to accept that we might never see each other again ... It tore me apart.'

She overhauled her beloved BMW bike and packed away the other mementoes and journals of her round-the-world ride. If the world didn't want to know about her epic circumnavigation, perhaps she should just consign it to personal memory.

## BUILDING A NEW LIFE

Beard and Mark resumed their relationship; in 1990 they had a son, Tom, before separating the following year, as Beard struggled with postnatal depression and the demands of converting a water tower near London into a home for herself. But she endured, finishing the tower (she lives there still) and qualifying as an architect. Beard opened her own practice in 1998. The riding and other adventures continued (she gained her pilot's licence, too).

Then, over 20 years after her return, BMW published an article about her exploits on its website. The tale of Elspeth Beard's journey spread. Hollywood agents flew her to Los Angeles to discuss a possible film. She declined that offer, but the interest prompted her to dig out her boxes of memories.

'Writing the book really helped me sort things out in my head,' she recalled to Lonely Planet. 'I had never thought about my relationship with my parents. My whole relationship with Mark and having Tom. It was hard but it was brilliant.' Hardest of all was learning what became of Robert. It transpired that he had died in 2009. Subsequently, she learned from friends and family touching details of his deep feelings about her and their time on the road together, and the profound effect it had on the rest of Robert's life.

To read in depth about those revelations, turn to *Lone Rider*. Suffice it to say here, Beard's journey ultimately found the meaning it had long lacked: 'Thirty years after my trip had been forgotten by almost everyone, I knew that there was one very special person for whom it had always meant as much as it did for me. Now, at last, I knew that everything – accidents, illnesses, medical emergencies, hardships, heartbreaks, not to mention more than 56,325km (35,000 miles) and two years on the road – had, after all this time, been worth every wretched and every glorious minute.'

PIONEERS OF FEMALE ADVENTURE MOTORCYCLING

# ANNE-FRANCE DAUTHEVILLE

Bored with her advertising job in early '70s Paris, she ditched her career for the call of the road, and, packing plenty of style, became the first woman to ride around the world.

S tifled by what she saw as her conformist life in early 1970s Paris, the 28-year-old Anne-France Dautheville resigned from her job in an advertising agency, bought a 750cc motorcycle and in 1972 set off on with 100 other riders on the inaugural Raid Orion, a 7000km (4350-mile) ride from Paris to Isfahan, Iran. She was, she said, the only woman on the start line.

Her appetite whetted, the following year Dautheville mounted another motorcycle trip, this time solo, and one that would eventually take her round the world – making her, it's thought, the first woman to do so. On the 20,000km (12,420-mile) route, Dautheville travelled to Canada, Alaska, Japan, India Pakistan, Afghanistan, Iran, Turkey, Bulgaria, Yugoslavia, Hungary, Austria, Germany and then France. She undertook her journeys in style and was rightly aware that she had a dramatic story to sell – which she did, both writing journalism to pay her way, and then penning popular books such as *Une demoiselle sur une moto* after the Raid Orion and *Et j'ai suivi le vent* after her round-the-world trip.

'I am a normal woman,' she told the *New York Times* in 2016. 'I am not exceptional at all. I'm not especially courageous, I'm not especially strong ... it was the first time an average person could do all these things.' After a seven-month trip around South America, Dautheville gave up adventure riding in the early '80s, blaming the prosaic imposition by the French government of currency controls, and turned to the rather more sedate pastimes of novel-writing and gardening.

Clockwise from left: cooking in the Australian Outback in 1975; picking herself up in Iran in 1972; in a Kenzo dress after her 1973 circumnavigation; the 100cc Kawasaki that took her round the world, before one of the epic Buddha statues of Bamiyan, Afghanistan.

# BIG WHEELS, KEEP ON TURNING

**Ted Simon is the daddy of the RTW ride. Between 1973 and 1977 he experienced adventure and romance, revolution and enlightenment on his epic 101,000km trek.**

A modern jet airliner can circle the world in 48 hours. It took Ted Simon four years, but he put in every gritty, dusty mile. The decision to ride a motorcycle around the world took minutes; the physical journey – from London to Italy, Tunisia to South Africa, Brazil to California, Australia to Malaysia, India to London – became an odyssey, beset by every travel trauma imaginable, from mechanical breakdowns to being thrown in jail as a foreign spy. The remarkable thing was not how long the journey took, but the fact that it was completed at all.

## I QUIT

In 1973, the year Ted Simon packed in his job as a newspaper editor and set off on his 101,400km (63,000-mile) expedition around the world, travelling was both trickier and simpler than it is today. The cheap air travel revolution had yet to take place. Communications were limited to letters and landline phone calls. And overland travellers had to dodge a gauntlet of revolutions, civil wars and independence uprisings to cross the continents. On the flip side, in an age before every detail was recorded online, travellers could overcome all sorts of obstacles with the right balance of politeness and personal charm.

Critically, for an aspiring motorcyclist with ambitions to circle the globe, the motorcycles of the 1970s were manual, mechanical machines; most problems could be fixed with just a spanner, a piece of bent wire and some grease, elbow or otherwise. The 1970s were

also an age of innocence, at least in travel terms. Without 24-hour online information, would-be globetrotters were happy to set off armed with only sketchy information gleaned from pub conversations, confident that they would be able to draw on the global tradition of kindness and hospitality to strangers in the event of any trouble.

## PASSING THE TEST

Nevertheless, making the decision to pack his whole life onto the back of a British-built motorcycle and set off for an adventure of undetermined duration was a big deal for Ted Simon. Among other things, he wasn't an experienced rider. The journey was almost cancelled before it began, when the wannabe circumnavigator failed his first motorcycle riding test; he passed on the second attempt, just in time to set off on his journey. At the time, Simon had no idea that the journey that would later be recorded in his bestselling travelogue, *Jupiter's Travels*, would become the defining event of his life.

With limited baggage space, packing became a fine balancing act between preparing for every eventuality and fitting everything into the panniers and storage boxes that could be strapped onto a Triumph Tiger 100 500cc. Simon spent six months transforming his motorcycle into a mobile home on two wheels, whittling his kit down to the travel essentials – a camera and film, a stove and water bottle, a tent and mosquito net, a passport and money, a medical kit, a warm flying jacket, a formal coat for potential embassy parties, and a pared-down toolkit to keep his trusty mechanical steed on the road.

Ted Simon and his fully laden Triumph Tiger 100 in the Argentine hinterland.

© TED SIMON

But for the former London hack, simplicity was rather the idea. Simon saw his trip around the world as a philosophical as much as a physical journey, an experiment to see what might be possible without the material and social constraints that prevent people from doing exactly what they want to do at any point in time, stripped of the comfort blanket of convenience that holds most of us back from living fully in the moment.

## HOW TO RIDE IN THE DESERT

Simon's four-year extravaganza started prosaically enough, in late 1973, with a comfortable ride on the well-surfaced highways of Europe, from London to the heel of Italy, to pick up the ferry to Tunis. For the novice motorcyclist, this was a chance to learn his machine – its balance, its mechanical idiosyncrasies, and the unique soundtrack of clangs and vibrations which could indicate anything from a loose strap to all-out disaster. It was a training slope, so to speak, for the mountain peak that was to follow.

Simon's route through Africa was circuitous. With the Yom Kippur War sealing off passage from Israel into Egypt, his best option was to enter North Africa via Tunisia and take a chance on rarely used border crossings between Tunisia, Libya and Egypt. In a pre-internet age, much had to be taken on trust – the route was planned based on chats with fellow travellers, addresses provided by friends of friends, and roads marked as dotted lines on decades-old maps.

Leaving Tunis, Simon had barely reached the desert before the first of many falls. The trick to desert riding, it transpired, was not to avoid falling, but to fall with minimum damage to yourself and the bike. The soft sand helped. In the event, Simon's winding route was notable for unprecedented rainfall and sections of the desert were flooded.

Crossing into Libya was the easy part. Entering Egypt, a nation locked down by conflict, was a far more challenging proposition. Despite being repeatedly told that he had no chance of crossing the border between Libya and Egypt – and having a visa that specifically excluded crossing via the coast – Simon decided to chance it, riding the 1000km (620 miles) from Tripoli to El Salloum and wrangling his way across the border with a mixture of charm, politeness and good luck.

## HOW NOT TO RIDE IN THE DESERT

Once in Egypt, the traverse of Africa could begin in earnest. Not for the first time, however, the road took its toll. On a night-time charge towards Marsa Matruh for fuel, Simon lost the lid to one of his storage boxes, and with it, his wallet, passport and travel papers. Amazingly, by combing back along the route, the passport and documents were found; the money and credit cards were not.

A less dedicated rider might have reconsidered the trip, but Simon pushed on to Alexandria, where he performed the first of many major pieces of surgery on his temperamental, British-built machine, rebuilding pistons deformed by running hard in the desert heat. Alexandria was also the location for the first of many arrests, after local people became convinced that Simon was an Israeli spy.

Again, charm – and a cutting from the *Sunday Times* announcing his journey – got Simon out of a tight spot. He was freed to continue his journey, but not by road, thanks to troop movements along the main Nile highway. Instead, bike and rider were loaded onto the train from Cairo to dusty Aswan, where Simon was able to resume his ride, far from Egypt's troubled northeastern border.

Crossing Lake Nasser on a decrepit ferry, Simon came up against more bureaucracy in the border town of Wadi Halfa. Police insisted that the journey continue by train as far as Atbara, where Simon was finally able to gun the throttle and clatter through the deserts of northern Sudan, at the time enjoying a 10-year hiatus in its long civil war.

## AFRICA RISING

Road travel through sub-Saharan Africa in the early 1970s involved daisy-chaining between widely spaced petrol pumps, breaking the journey at rough-and-ready roadhouses frequented by traders, mercenaries, thieves, prostitutes and old colonials, washed up by time and circumstances on the biggest beach in the world. At most stops along the way, the bike had to be painstakingly restored to roadworthiness in preparation for the next leg of the journey.

Any semblance of a proper road vanished in Sudan. Reaching Kassala involved tracing the tracks left by freight trucks in the sand, using the Atbara River as a compass, carefully navigating a passage between sand that was either too dry or too wet to ride on. Maintaining supplies of petrol and water became a life-or-death necessity, requiring multi-day detours and strange, often humbling encounters with desert dwellers.

The long journey on to Meterna on the Ethiopian border provided plenty of time for contemplation, and some of the most challenging riding conditions on Earth. It was only on the final approach to Gondar, just north of Lake Tana, that tarmac reappeared. (➔ p.273)

Left, from top: encounter with a muddy puddle on the road to Benghazi, Libya; visiting the ancient Roman ruins of Cyrene, near the present-day Libyan city of Shahhat.

## DEATH OF THE HIPPIE TRAIL

For three decades, a procession of beatniks and then hippies travelled from western Europe, through Turkey, Iran, Afghanistan, Pakistan, India and on to Southeast Asia. The first Lonely Planet book, *Across Asia on the Cheap* (1973), catered to that market. But political upheaval – the Yom Kippur War, the Lebanese Civil War, the Iranian revolution and the Soviet invasion of Afghanistan – in the 1970s stemmed the flow of travellers.

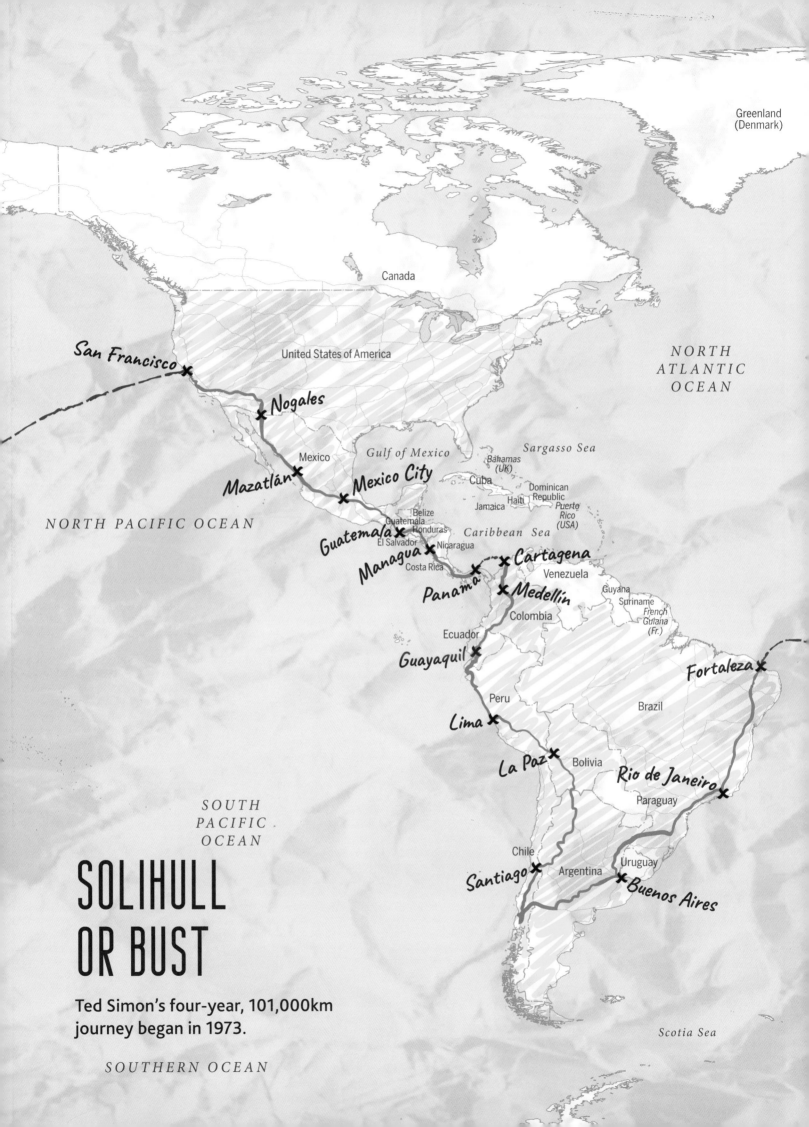

# SOLIHULL OR BUST

Ted Simon's four-year, 101,000km journey began in 1973.

# RIDING THE WORLD

Long or short, quick or slow – here are just a few
of the notable motorcycle circumnavigations.

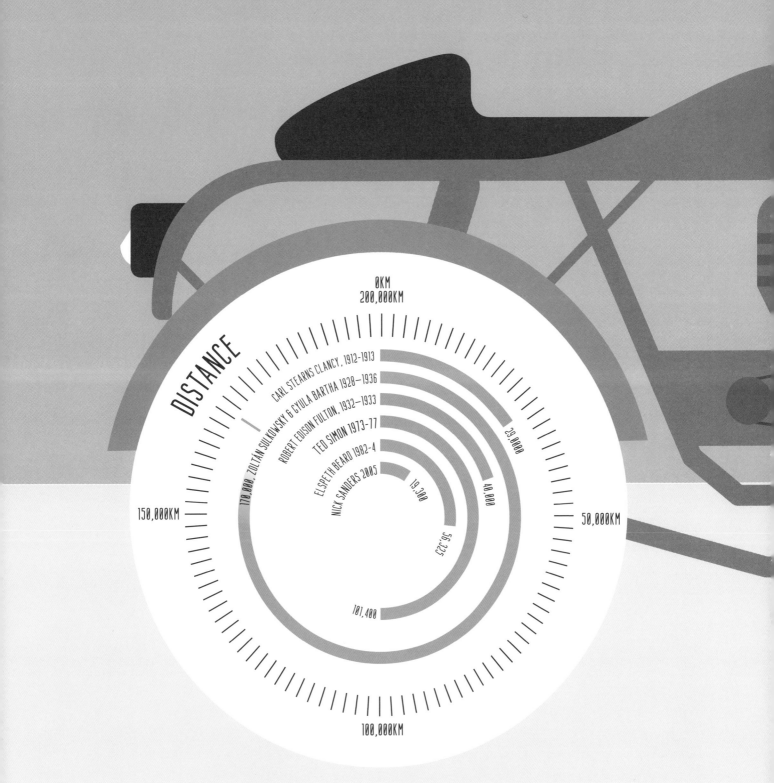

DISTANCE

0KM
200,000KM

CARL STEARNS CLANCY , 1912–1913

170,000, ZOLTÁN SULKOWSKY & GYULA BARTHA 1928—1936

ROBERT EDISON FULTON, 1932—1933

TED SIMON 1973–77

ELSPETH BEARD 1982–4

NICK SANDERS 2005

20,0000

19,300

40,000

56,325

101,400

50,000KM

150,000KM

100,000KM

ARCTIC OCEAN

Greenland
Sea

Svalbard
(Norway)

Kara
Sea

Barents Sea

Norwegian Sea

Iceland

Norway
Sweden
Finland

North
Sea

United
Kingdom
Denmark

Union of Sovet Socialist Republics

Ireland

**London**
**Start 1973**
**End 1977**

Netherlands
Belgium
West
Germany
East
Germany
Poland

France
Switzerland
Austria Hungary
Czechoslovakia

Romania

**Montpellier**

Italy
Yugoslavia
Bulgaria
Albania

**Istanbul**

Caspian
Sea

**Kabul**

**Peshawar**

Portugal
Spain

**Palermo**

Greece
Turkey

**Tehran**

Afghanistan

**Kathmand**

**Tunis**

Tunisia

Cyprus
Syria
Lebanon
Israel Jordan
Iraq

Iran

Pakistan

**Delhi**

Nepal
Bhutan

**Benghazi**

Kuwait

Morocco

**Alexandria**

Saudi
Arabia

Qatar

Bangladesh

Libya

Egypt

UAE

**Calcu**

Western
Sahara

Sahara Desert

**Wadi Halfa**

**Bombay**

India

**Puri**

Bu

Mauritania

Mali
Niger
Chad

Oman

Arabian Sea

**Atbara**

Sudan

North
Yemen
South
Yemen

**Madras**

Senegal
The Gambia
Guinea
Bissau

Upper Volta

**Gondar**

Djibouti

**Cochin**

**Trincomalee**

Sierra
Leone

Guinea

Ivory
Coast

Benin
Togo

Nigeria

Central African
Republic

Ethiopia

Somalia

Sri
Lanka

Maldives

Liberia

Ghana

Cameroon

Equatorial
Guinea

**Moyale**

Gabon

Congo

Zaire

Uganda

Kenya

Rwanda
Burundi

**Nairobi**

Sin

**Mombasa**

Tanzania

Seychelles

**Mbeya**

**Ndola**

Angola

Malawi

Comoros &
Mayotte
(Fr.)

INDIAN
OCEAN

**Livingstone**

Zambia

Mozambique

South-West
Africa
(S. Afr.)

Rhodesia

**Mutare**

Madagascar

Botswana

**Johannesburg**

**Lourenço Marques**

Swaziland

South
Africa

Lesotho

SOUTH
TLANTIC
OCEAN

**Cape Town**

**Port Elizabeth**

TIME

0 YEARS
10 YEARS

10 MONTHS

18 MONTHS

26 MONTHS

4 YEARS

CARL STEARNS CLANCY , 1912–1913

8 YEARS, ZOLTÁN SULKOWSKY & GYULA BARTHA 1928–1936

ROBERT EDISON FULTON , 1932–1933

TED SIMON 1973–77

ELSPETH BEARD 1982–4

NICK SANDERS 2005    19 DAYS

7.5
YEARS

2.5
YEARS

5 YEARS

But the ride through Ethiopia was not a happy one; under Haile Selassie's autocratic rule, rebellion and famine stalked the country.

It was with some relief that Simon left Ethiopia by way of Moyale, following a route that would later become a favourite with overland truck tours. Despite a decade of independence, Kenya carried lingering traces of British colonialism. Simon dropped in on white farmers preparing for the day when their farmland would be taken over by the Kenyan government, and sipped cold Tusker beers with Kenyans who were itching at the chance to build themselves a new African identity.

In the first of many detours, Simon hitched a ride on a plane to Lodwar on Lake Turkana, before diligently overhauling his motorcycle for two weeks in preparation for the ride to Mombasa and the coast. A flat tyre and damaged rim 240km (150 miles) out of Nairobi put the dampener on his onward plans until a replacement could be delivered from the capital, meaning days in the company of rogues and road traders at Kibwezi.

## DOCTOR BIKE

Riding south from Mombasa towards the Tanzanian border, Simon experienced his first wave of loneliness and depression. At many points in the trip, this lapse of morale would arrive without warning – usually around five o'clock in the afternoon – but if resisted, it would fade just as quickly. Crossing into Tanzania at Lunga Lunga, Simon picked up the Tanzam Highway at Dar es Salaam, encountering zebras, giraffes, baboons and other wildlife as regular companions – and often obstacles – on the road.

The trip on through Zambia passed smoothly, but Rhodesia (soon to become Zimbabwe) was still living under the intolerant rule of Ian Smith – outwardly a prosperous, white nation, but with an impoverished black majority. The same, demoralising sense of conspicuous inequality followed the writer to South Africa, where a smooth run to Johannesburg was halted by disaster. A piston, damaged by heat back in Egypt, finally shattered, filling the engine with shrapnel.

Simon limped on to Naboomspruit and spent two days rebuilding the engine. With emergency stops for further repairs, he crept, snail-like into Jo'burg, pausing for some well-earned rest and recreation with acquaintances. Here fate once again interceded on the route – the oil crisis caused by the Yom Kippur War scuppered any chances of sailing from Cape Town to Rio de Janeiro, but a new possibility opened up: crossing by slow freighter from Lourenço Marques (later renamed Maputo) in Mozambique to the Brazilian port of Fortaleza.

There was a satisfying continuity to the route – leaving Africa's most important Portuguese colony and sailing for the largest Portuguese colony in the Americas. Simon jumped at the opportunity, taking a grand tour through Cape Town and the historic Dutch towns of the Garden Route to Durban and Swaziland, arriving in Lourenço Marques just in time for the start of Mozambique's independence revolution.

## THE SPY WHO DIDN'T LOVE ME

Simon wasn't entirely sad to leave Africa's complex conflicts and board the rusting *Zoe G* for the long sailing across the Atlantic, but had he known what was waiting for him, he might have been less enthusiastic. On arrival in Fortaleza, the writer was held up by

Above: sunset scene during Simon's ferry crossing of Lake Nasser, Egypt; like most of the ferry passengers, Simon gave the second-class deck a wide berth, preferring to sleep on the roof.

irregularities over his motorcycle importation papers and then promptly arrested as a spy. When he was released two weeks later, the absurdity of Simon's situation was compounded as one of the female officers responsible for his imprisonment attempted to initiate a love affair.

Simon fled on through Brazil, riding first to Rio, where his list of contacts opened doors to senior figures in Brazilian high society, including former president Juscelino Kubitschek. But before long the roaming resumed, and Simon gunned his trusty Triumph towards Foz do Iguaçu, choosing on a whim to cross to Argentina rather than Paraguay. Sewing extra insulation into his flying jacket, the road-hardened traveller began the slow, arduous process of zigzagging back and forth across the Andes through Chile and into Peru, in the company of some fellow free spirits who were driving a battered van across South America.

## EASY COME, EASY GO

Somewhere along the way, Simon achieved acceptance of the privations of travelling by motorcycle. Comfort, where it was a possibility, was dictated by the condition of the road. Accommodation was dictated by range. If no

## LAP OF HONOUR

In 2001, aged 70, Ted Simon repeated his 1973 trip on a BMW R80GS, doing so in just three years. On the way, he met actor Ewan McGregor and TV presenter Charley Boorman in Mongolia, in the middle of their own motorcycle odyssey from London to New York – a trip inspired directly by *Jupiter's Travels*.

town was reached by the time darkness fell, or petrol ran low, or bad weather descended, bed was a sheet by the roadside.

Going without bread, or meat, or milk, or beer, became commonplace. At times, Simon fished, or haggled or begged for fish, on remote Pacific beaches. At other times, he hunted for crabs with a purse-sized revolver. Large outposts of civilisation were avoided as a precaution, after the government in Lima was toppled in yet another military coup.

Travel on two wheels, however, provided ample reward: the sensation of being there, wherever 'there' happened to be at any given point in time, rather than watching the world pass by like a spectator through the window of a bus or train. The most rewarding experiences were not sights, or landscapes, or even the mechanical thrills of riding, but interactions with ordinary people, of the kind that would be impossible travelling in any other type of motor vehicle.

## INTO ESCOBAR'S COLOMBIA

Simon continued with his van-driving companion, Bruno, through the relative calm of Ecuador into the lawlessness of 1970s Colombia, where a young Pablo Escobar was beginning his ascent to the top of the Medellín Cartel. The van crashed near Popayán, and the pair went their separate ways; Simon rode as far as the coast at Cartagena, then flew himself and the bike to Panama to avoid the impassable Darien Gap, the only break in the 30,000km (19,000-mile) Pan-American Highway.

On the trip up through Central America, Simon met other riders with battle tales to tell. He crossed into Costa Rica, zipped through Nicaragua, and detoured to Mayan ruins in (then peaceful) Honduras and Guatemala. The faithful flying jacket was lost on a rough stretch of road in Mexico; with some tinkering and roadside repairs, the bike limped on to the US border at Nogales. Simon entered the United States 20 months and 40,200km (25,000 miles) after first turning the key in the ignition in London.

## THE ROAD OR LOVE?

Rolling on crisp, smooth tarmac through Tucson and Phoenix, Simon pointed his ailing motorcycle towards Los Angeles and the US offices of the Triumph Motor Company, where he was received with some bewilderment. Looking like a wild man after months exposed to the elements, Simon was somehow able to talk his way into a comfortable hotel room while Triumph engineers put his venerable machine back on the road.

At around the same time, the writer looked up an address passed on by some motorcyclists he had encountered in Quito, and reached a crossroads. He moved into what could loosely be described as a commune, fell in love, and postponed his journey on to Australia. But

despite the allure of barefoot country living, it wasn't long before the call of the road resumed its siren song. Duly, Simon saddled up the Triumph, said his goodbyes – promising to return once the journey was completed – and hit the highway.

Arriving in Sydney off the boat from San Francisco, the writer had originally planned to ride to Darwin, and from there sail to Indonesia. So Simon rode north up the east coast to Cape Tribulation, Queensland, through floods at Lotus Creek, where he spent time with long-distance truck drivers stranded by the rainy season.

But the military invasion of Timor and the cyclone that ripped through Darwin necessitated a new route. Turning inland, he rode south to Melbourne. There, he worked on the Triumph and secured space for himself and the bike on a ship bound for Singapore.

## FROM PERTH TO PENANG

The Triumph now ready for this ride across southern Australia, Simon beach-hopped along the coast between Melbourne and Adelaide, before entering the acacia-dotted expanse of the Nullarbor Plain – 1200km (745 miles) of ruler-straight roads beset by treacherous hazards such as hidden dust pockets and heavy-as-a-horse red kangaroos leaping unexpectedly onto the highway.

Arriving rattled but invigorated in Perth, Simon boarded the ship with a flock of sheep (and a waspish Welsh captain) and came down with a flu on the voyage to Singapore. Having disembarked, he recuperated in Malaysia's Cameron Highlands. There, he wrote a letter excusing himself from returning to his love affair

Below, from left: custom-made panniers under construction in Villaguay, Argentina; sweeping views during the South American leg of Simon's epic ride. Opposite: a break from the bike en route to Wadi Halfa, Sudan – helmet required.

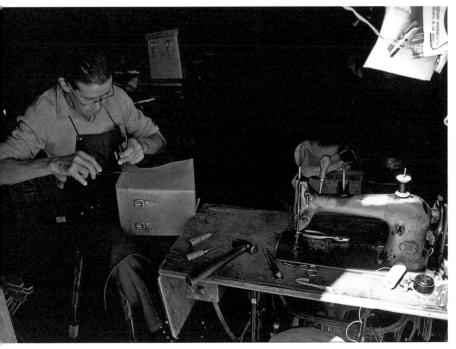

© TED SIMON

in California. As if in punishment for this act of faithlessness, the bike failed again on the ride to Penang, forcing the writer to hole up for weeks waiting for parts to be shipped to Malaysia.

At the time, Penang was a feverish port city, rife with prostitution, opium smuggling and organised crime, and Simon passed the time as a fascinated observer, wandering the

Above: camping by the roadside in Turkey. Below: sightseeing during a pause for bike maintenance in Bombay (Mumbai), India.

shophouse-lined backstreets and fishing from the esplanade, before being hit in the eye by a flying fishhook. While hospitalised, his wallet, passport and list of contact details of friends of friends were stolen.

Getting a new passport meant looping back to Kuala Lumpur for a tortuous dose of British bureaucracy, but eventually Simon was able to secure passage to Port Blair in the Andaman Islands, and on to Madras in India (the obvious overland route via Thailand, Burma and Bangladesh being blocked by the Communist government in Rangoon). If ever there was a country made for motorcycling, India was it.

### THROTTLING BACK IN INDIA

Staying with Indian friends eased the transition into life on the subcontinent, before another crisis struck – Simon's stepfather died unexpectedly, and he locked up his motorcycle and flew home for the funeral. Returning to India, Simon made up his mind to find something of spiritual value in his journey, and joined the pilgrims thronging the temple trail through Kanchipuram, Mahabalipuram (Mamallapuram) and Tiruchirappalli.

This voyage of self-discovery traversed half

# BY MOTORCYCLE

of a replacement from the UK. Meanwhile, storm clouds were gathering in Pakistan with the collapse of the Bhutto regime and the imposition of martial law.

But luck was on Simon's side. With the new republican government holding tenuous control over Afghanistan, he was able to cross into Pakistan and follow the Khyber Pass to Kabul, then a drug-fuelled hang-out on the Hippie Trail. Covering the final 9700km (6000 miles) involved long rides between outposts of civilisation in remote parts of Iran and Turkey, but desert riding held few fears for the now-veteran motorcyclist; he was more worried by the prospect of ending his journey and adjusting to life back at home.

## SOLIHULL OR BUST

The final stages of the trip meandered through Greece and followed Yugoslavia's notorious coastal highway into the heart of Europe. Within four days, Simon had reached Munich; another day brought Switzerland into view. In June 1977, Simon reached his home in France, and offloaded his small collection of souvenirs from four years on the road – trinkets, spices, herbs, a teapot, a carpet, a Russian samovar – before riding on to London to close the circle at the Meriden Triumph Motorcycle works near Solihull, where his motorcycle began its life four years and 101,400km (63,000 miles) earlier.

© TED SIMON

**Below: riding through the desert toward Lodwar, Kenya.**

the Indian subcontinent. He had audiences with temple priests in Tiruvannamalai and Madurai and sought the silence of the hills in Ooty and Kodaikanal. He chased down the conjuring guru Sai Baba (later disgraced by sexual misconduct allegations) at his Bangalore ashram. For a while, the writer even entertained the thought of becoming a guru himself. Simon crossed to Sri Lanka and progressed through Colombo, Kandy and Trincomalee.

Over many months, Simon drank deeply of the subcontinent. He kicked back on beaches in Goa. He became a house guest of the Maharaja of Baroda. He followed the Hippie Trail from Delhi to Kathmandu. He stayed at the iconic Salvation Army Hostel on Sudder Street in Calcutta, and photographed erotic statues at Khajuraho. And he kept exploring, following the Brahmaputra River all the way to the Burmese border in Assam.

## BACK TO REALITY

The last leg of the journey – from India across Pakistan, Iran, Turkey and Europe – was held up again by mechanical worries. A broken chain revealed a sprocket too worn to safely ride on, and weeks were lost in Delhi waiting for delivery

© TED SIMON

# INDEX

# INDEX

# ACKNOWLEDGEMENTS

**Front cover images:** © Rob Wilson, OSTILL is Franck Camhi, Sailorr / Shutterstock, © Topical Press Agency, Bettmann, Pool, izhairguns / Getty Images, Globe image courtesy of Bellerby & Co Globemakers / bellerbyandco.com

**Back cover images:** © Elspeth Beard, © Iakov Filimonov, lily_of_the_valley, saiko3p, Praneet Soontronront / Shutterstock, © Elena-studio / Shutterstock, © BikingMan / David Styv, © james Robertson

**p10-11 Ship Illustration:**
© Sylvain Sonnet, Victor Fraile Rodriguez, CRISTINA QUICLER, UniversalImagesGroup, Samohin / Getty Images, © Elenarts, Sailorr / Shutterstock

**p40-41 Bicycle Illustration:**
© Fox Photos, ispyfriend / Getty Images, © TTstudio / Shutterstock. © BikingMan / David Styv, © james Robertson

**p72-73 Plane Illustration:**
© Topical Press Agency, PhotoQuest, New York Daily News Archive, Bettmann / Getty Images, © Mark Brodkin Photography, Skadr / Shutterstock

**p100-101 Boat Illustration:**
© Terrence Spencer, Dhammika Heenpella / Getty Images

**p120-123 Donald Crowhurst Illustration:**
© Eric Tall / Keystone, Sean Gladwell / Getty Images,

**p125 Illustration:**
© Michel Porro / Getty Images

**p133 Foot Illustration:**
© David Crump / ANL, MicheleB Shutterstock, © Lynn Pelham, Paul Harris / Getty Images

**p134 Illustration:**
© Scheufler Collection / Getty Images, © Philip Lee Harvey / Lonely Planet, © Suthin _Saenontad / Shutterstock

**p136 Illustration:**
© Paul Harris, jejim, SeanPavonePhoto, Fox Photos / Getty Images, © Diego Grandi / Shutterstock

**p154 Illustration:**
Globe image courtesy of Bellerby & Co Globemakers / bellerbyandco.com © Margarita Solianova / Getty Images

**p144 Illustration:**
© Paul Harris, Fox Photos / Getty Images, © Back Page Images, diy13 / Shutterstock

**p161 Illustration:**
© ullstein bild / Getty Images

**p186-187 Balloon Illustration:**
© ullstein bild Dtl. Tim Ockenden - PA Images, August Bocker / Getty Images, EDB Image Archive / Alamy Stock Photo

**p214-215 Train Illustration:**
© Amir Ghasemi / Getty Images, © Museum of the City of New York/Byron Collection / Getty Images, © saiko3p / Shutterstock, © Matt Munro / Lonely Planet, © JeniFoto, Praneet Soontronront / Shutterstock

**p248-249 Motorcycle Illustration:**
© Jonathan Stokes, Justin Foulkes / lonely Planet, © Elena-studio / Shutterstock, © mazzzur / Getty Images, © Elspeth Beard

First Edition
Published in October 2020
Lonely Planet Global Limited
CRN 554153
www.lonelyplanet.com
ISBN 978 17886 8937 3
© Lonely Planet 2020
Printed in Malaysia
10 9 8 7 6 5 4 3 2 1

Managing Director, Publishing Piers Pickard
Associate Publisher & Commissioning Editor Robin Barton
Art Director Daniel Di Paolo
Cartography Wayne Murphy
Editors Monica Woods, Mike Higgins
Proofer Cliff Wilkinson
Indexer Polly Thomas
Picture Research Katy Murenu
Print Production Nigel Longuet
Thanks to Flora MacQueen

Written by:
Mark Mackenzie: Ship, Bicycle (with Jenny Graham by Robin Barton), Plane, Foot,
Balloon and Boat (with Navigation 101 by Roger Barton, RN)
Joe Bindloss: Train, Motorcycle (Ted Simon). Oliver Berry: Car.
Peter Thoeming: Motorcycle (Robert Edison Fulton and first-person story).
Mike Higgins: Motorcycle (Elspeth Beard and Pioneering Female Adventurers)

Lonely Planet offices

AUSTRALIA
The Malt Store, Level 3, 551 Swanston Street, Carlton Victoria 3053 Phone 03 8379 8000

IRELAND
Digital Depot, Roe Lane (off Thomas St), Digital Hub, Dublin 8, D08 TCV4

USA
Suite 208, 155 Filbert St, Oakland, CA 94607 Phone 510 250 6400

UNITED KINGDOM
240 Blackfriars Road, London SE1 8NW Phone 020 3771 5100

STAY IN TOUCH
lonelyplanet.com/contact